Toward a History of Game Theory

Toward a History of Game Theory

Annual Supplement to Volume 24

History of Political Economy

Edited by E. Roy Weintraub

Duke University Press

Durham and London 1992

Printed in the United States of America
on acid-free paper ∞
ISBN 0-8223-1253-0
This is the 1992 annual supplement to
History of Political Economy, ISSN 0018-2702.

Contents

Part 3 Crossing Disciplinary Boundaries

Toward a History of Game Theory

Introduction

E. Roy Weintraub

If one looks back to the 1930s from the present and reads in the major economic journals and examines the major treatises, one is struck by a sense of "the foreign." That past period, notwithstanding G. L. S. Shackle's (1967) claim that it was a decade of high theory, produced works that today require a gloss, a modernization of the text as it were. Keynes's work, that of Hayek, that of the business cycle theorists seem to be written in a primitive language particularly unsuited, we feel today, to the problems they addressed.

If, on the other hand, we read economics journal articles published in the 1950s, we are on comfortable terrain: the land is familiar, the language seems sensible and appropriate. Something happened in the decade of the 1940s; during those years economics was transformed from a "historical" discipline to a "mathematical" one. To be sure, there had been past calls for the mathematization of economics and attempts to do economics mathematically. From Jevons in England, Walras in France, and Pareto in Italy those calls were clear and forceful. From the American Irving Fisher they were more pointed still. Yet the mathematical movement was slow to take root in the mainstream economics discourse. The founding of the Econometric Society in 1930 was a signal that times were more hospitable to a mathematical approach to economics, but the mathematical economists of the 1930s were still, in Herbert Simon's words, "a sect" (Simon 1959, 493).

Although economists have been interested in this transformation of economics and have recorded personal memories of this period in autobiographical notes, Nobel Prize speeches, and so on, it is only recently that historians of economics have begun to take an interest in the matter. Perhaps this is because the period is a relatively recent one, or perhaps it is because the issues have a very technical context or are large and confusedly interdisciplinary. Yet the past several years have produced some attempts to write the history of this period; individual scholars have taken on the challenge of tracking the mathematization of modern economics. Recent books by Philip Mirowski (1989) and Bruna Ingrao and Giorgio Israel (1990) have begun this work, telling the complex and interrelated tales of mathematics, physics, and economics in the first several decades of this century. This history is hard to write because it covers many disciplines and many countries. For example, a history of existence of equilibrium proofs over the period 1930 to 1954 (as in Weintraub 1985; Punzo 1991) must connect the Vienna of the mathematician Karl Menger, the "Mathematische Kolloquium," to John von Neumann (1936) and Abraham Wald and trace the movement of ideas and people from Europe to the United States in the 1930s. The story links the Cowles Foundation and the Econometric Society with Kenneth Arrow, Lionel McKenzie, and Gerard Debreu, and thereby links the developments in game theory and programming to the emerging general equilibrium literature.

This line of history is only loosely connected with the issues of how to model economic dynamics. These issues, which greatly concerned economists in the 1930s, found their interpreter in Paul Samuelson (1947). The confused and multilayered history of how "dynamics" circa 1930 was transformed into "stability theory" in the economics literature by the 1950s is likewise hard to tell in terms of "The Story of How Economist A Influences Economist B." Leaving economics aside, one excellent model for this kind of history can be found in Russett 1966. In economics, the dynamics story connects the mathematical community—J. W. Gibbs, E. B. Wilson, Poincaré, Liapunov, Andronov, G. D. Birkhoff, Solomon Lefschetz—with Samuelson, Arrow, and Hurwicz. Specifically, the history is one of the reception in economics of certain themes, results, and approaches in the study of dissipative dynamical systems and the related qualitative theory of differential equations. The narrative line involves the "Harvard Pareto Circle" of physiologist L. J. Henderson (Henderson 1935; Heyl 1968) and the unusual configuration of skills and talents of the Princeton mathematician Solomon Lefschetz. The story line should indicate that the new ideas of equilibrium theory were not unique to economics but rather were "in the air" so to speak, becoming manifest in a variety of fields and disciplines from sociology to psychology, from naval ballistics to development economics (Weintraub 1991).

Of course there was another major transformer of economics in the

1940s, an intellectual upheaval in the sciences which has been termed "the Probabilistic Revolution" (Kruger et al. 1987; Kruger, Gigerenzer et al. 1987). In economics, this became the new econometrics signaled by Haavelmo's work and that of the Cowles Foundation econometricians, but linked to earlier works in statistical economics in England, Holland, and the United States. Mary Morgan (1990) presents a wonderfully rich history of this diffusion of statistics, economics, and mathematics.

Thus there are a number of perspectives that one could have and that could shape one's writing of a history of the 1940s. By noting the various changes that occurred in the way economists presented their analyses and formed themselves into a scientific community in the postwar period, one becomes interested in reconstructing the transformation of discursive practice in economics in the 1940s. But from all these points of view one is led again, and most seriously, to von Neumann and Morgenstern's *Theory of Games and Economic Behavior* (1944) which had a most profound and complex effect on the mathematization of economics in the postwar period.

Moreover, in the 1980s and now in the 1990s, game theory is perhaps the most lively field of economic theory, and game theorists are currently rewriting entire parts of what had been considered settled fields (e.g., Kreps 1990). This fact, too, piques the interest of a historian of economics in the background and development of the theory and provides a second reason for wishing to have a history of the development of game theory.

Finally, the history of the mathematics community has been more concerned with traditional mathematics, which mostly means the various subdisciplines of algebra, analysis, and geometry. More recently, historiographic changes have been driving new approaches to the history of mathematics, as that subdiscipline has shared the movement to social and contextual history of science.

> This historiography measures events of the past against the standards of their time, not against the mathematical practices of today. The focus is on understanding the thought of the period, independent of whether it is right or wrong by today's account. The historiography is more philosophically sensitive in its understanding of the nature of mathematical truth and rigor, and it recognizes that these concepts have not remained invariant over time. This new historiography requires an investigation of a richer body of published and unpublished

> sources. It does not focus so exclusively on the great mathematicians of an era, but considers the work produced by the journeymen of mathematics and related scientific disciplines. It also investigates the social roots of mathematics: the research programs of institutions and nations; the impact of mathematical patronage; professionalization through societies, journals, education, and employment; and how these and other social factors shape the form and content of mathematical ideas. (Aspray and Kitchner 1988, 24–25)

But this reemphasis in historical studies of mathematics has not so far been directed outside traditional mathematical subject matter: "Applied mathematics, non-Euclidean and projective geometry, operations research, probability, and statistics have received little attention" (30). Thus an additional reason to focus the attention of historians on game theory is that this applied field allows the new historians of mathematics to have access to a scholarly gold field, outside the usual mathematical domain, in which many of the new historiographic themes may be employed to yield riches.

As is not uncommon in historical research, the local and contingent played a role in developing the history of game theory project, the results of which are collected in this volume. First, there is the simple fact that we have access today to the history of game theory through the memories of some quite active researchers such as Howard Raiffa and Martin Shubik, and that Shubik particularly was interested in having an examination of the historical issues. Second, through the generosity of Dorothy Thomas Morgenstern, the Economist Papers Project, Perkins Library, Duke University received the papers of her late husband Oskar Morgenstern. This meant that, together with the Library of Congress's holding of the von Neumann Papers, archival material of importance to the historians of game theory was available for study.

As a result, the economics department of Duke University, with some financial support from both the Duke Endowment and the John M. Olin Foundation, was able to host a conference on the history of game theory in October 1990. The papers presented at that conference were both invited and submitted, with the invitations generated by my conversations with Robert Bates, James Friedman, Alvin Roth, William Aspray, and Martin Shubik. During a year as a Fellow at the National Humanities Center in 1988–89 I spoke with and corresponded with a number of people who "signed on" to this project and whose papers are contained in this collection. But it quickly became apparent that the history that would emerge would be less a finished product, more a prolegomena to any future history. This was so for several reasons.

First, note that the conventional view of the history of game theory in economics is relatively simple to narrate. It was that von Neumann wrote a paper in the late 1920s on two-person games and minimax. Borel claimed priority but this claim was rejected as mistaken. Then von Neumann and Morgenstern got together in Princeton, wrote their book in 1944, and the word went forth. The story goes on to tell us that, unfortunately, economists were slow to see the importance of the theory. Thus although two-person theory was solved early on, the interesting issues became those of *n*-person cooperative theory. These problems took a long time to solve, but finally the core emerged as a good solution idea for economics and helped to unify topics in general equilibrium theory. Subsequently information problems emerged in microeconomic theory which could be studied by attention to the nature of the Nash equilibrium theory of noncooperative games, and this is the area of current work in game theory in economics. As the papers in this volume will make clear, this potted history is misleading in all its details.

Second, the conference showed that there was a great deal of confusion about the origins of the theory. Not only is there a real "priority" debate, but there is a great deal of misunderstanding about the nature and issues of the von Neumann collaboration with Morgenstern. From the several sessions it became clear that there was a variety of possible interpretations that could be convincingly presented, and defended, about the pre-1944 history.

Third, there are problems that the conference papers turned up about the fact that game theory was not very well received among economists in the 1940s. In the immediate postwar period game theory was viewed with some suspicion: it was not *really* economics. It received more attention in other disciplines, and thus the conference revealed that the history of its reception required attention to a more generalized social science perspective, as well as to the issues of the development of a new theory in "applied" mathematics.

And finally, the conference participants were acutely aware that the present importance of the theory in economics leads to reading current concerns into the past record, a natural Whiggishness which, because the theory seems to a modern game theorist to have always meant what it means today, confuses the past with the present. Although there is no record of this theme in this volume, it was reflected in the comments that enlivened the various sessions from those in attendance who were game theorists in departments of economics, mathematics, and political science and in schools of business.

As a consequence of these problems with constructing a full and coherent account of the history of game theory, the project evolved into this volume with its more modest claim to take those interested "toward a history of game theory." The papers are thus not definitive, but are rather more preliminary. With few exceptions, they were not extensively rewritten to take account of the other papers in the collection, and thus author A's narrative reconstruction may not take full cognizance of author B's archival, or interview, "find."

The first set of papers presents the most interesting historiographic problems for the reconstruction of the history of the theory of games. They overlap in different ways in their concerns and emphases, yet they all take different approaches, and they reach conclusions that are not mutually consistent.

The paper by Robert Dimand and Mary Ann Dimand is, on the surface, a traditional paper in the history of economic thought. The authors backtrack from the theory of games circa 1944 to the von Neumann paper of 1928 to the Borel papers of 1921–27. Their argument concerns the priority claim asserted by, and generally "awarded" to, John von Neumann. The issues are fairly clear in this presentation, namely that Borel's contributions are underappreciated by historians and that von Neumann's priority must be reassessed. Against this particular claim we have Robert Leonard's paper which argues, from a perspective at least in part shaped by a view of scientific work as knowledge creation within a community, that there was no community in which game theory could have taken root before the 1940s, and thus there was no possibility that the theory could have been developed before that time. In linking the mathematical ideas with the communities of mathematicians and scientists, Leonard contextualizes a set of ideas in ways congenial to the best of the new historians of mathematics (Mehrtens et al. 1981). I note further that Leonard's paper, connected to a companion piece he has written on the RAND corporation (1991), uses a set of interviews that he conducted with a number of the original members of the game theory community.

Urs Rellstab's paper was written in response to some of the papers given at the Duke conference. Rellstab used the Morgenstern diaries to reconstruct the von Neumann–Morgenstern collaboration, and his paper corrects a number of misunderstandings about the nature and details of that joint endeavor. It is less a reapportionment of credit than it is an examination of the separate and distinct interests of the two authors of *The Theory of Games and Economic Behavior*. His story of the evolution of their joint project and the chronology he presents are reminders that later reconstructions, based on published retrospectives, may be less useful than contemporary records when they are available.

Andrew Schotter, who edited Morgenstern's *Collected Papers*, and who has a real claim to be Morgenstern's last "student," presents a word-picture of Morgenstern and his intellectual development. He is concerned with the issues of the von Neumann collaboration as well, and his conclusions differ somewhat from those of Rellstab. Although Schotter presents Morgenstern's Austrian background as context for his game theory work, he is more concerned with tracking the issues that concerned Morgenstern through later (post-1944) work in the theory of games, linking concerns that appeared to motivate the theory's creation with what has emerged in the various literatures.

The provocatively titled paper by Philip Mirowski connects the various themes of this first part. It may be best to read it together with its companion paper (1991) on the military connection with the theory of games in the 1940s and 1950s, already published, to see the large issues to which Mirowski wishes us to attend. His paper presents an unusually convincing argument which suggests that the interests of von Neumann, and Morgenstern, were connected to themes in their own intellectual histories. He suggests that von Neumann's response to Gödel's work led away from a concern with axiomatization, the "Hilbert Program" as it is termed, and toward the problems of what would become automata theory, and thus proof-strategies, or computational strategies, or strategic behavior more generally. Thus for Mirowski, the von Neumann of 1928 and the von Neumann of 1944 were doing two different kinds of work in game theory and cannot be well presented as having been engaged in a continuous line of thought. This argument clearly problematizes the issues raised by Dimand and Dimand about priority, and connects with Leonard's effort to track the relevant scientific communities. Mirowski's presentation of Morgenstern as an antineoclassical economist highlights the other deviant strand in the 1944 book and suggests that the revolutionary message of that work, coming from not one but two real critics of orthodox economics, made acceptance of the book a real problem for economists. An implication, of course, is that until the message of the book could be domesticated, as it were, the theory could not, or better, did not, receive a fair hearing.

Part 2 contains exemplars of what historians of mathematics call "reception theory." That is, the three papers all recount the particular ways that the new theory of games made its way into the literatures of the time and the particular mechanisms by which game theory became known among the relevant communities of scholars; they jointly tell the story of the transition of the theory from the first to the second generation. Angela O'Rand, a sociologist of science, tracks the networks and connections among the early social science users of the theory. Her study of the diffusion of ideas connects the game theory movement to the larger theme of the mathematization of the social sciences in the postwar period and shows how the theory became a wedge in opening up social science to mathematical tools. This, of course, is particularly ironic history since the 1944 book itself was written with a thinly disguised abhorrence of the mathematization of economics going on at the time.

The two papers by Howard Raiffa and Martin Shubik are important documents of the "second generation" of game theorists. Based on personal recollections of Michigan and Princeton in the late 1940s, they show how the message of the theory of games grew into consciousness among the new generation of social scientists trained in the postwar period. The new theory was carried to those places, and those intellectual communities gave a hearing to the new ideas and of course transformed them as well. As personal testaments they will be useful to scholars interested in knowing how game theory captivated those whom we now consider to be the theory's "giants."

Finally, the three papers in part 3 concern the transmission of the new game-theoretic ideas into areas, and communities, apart from mathematics and economics. How ideas permeate disciplinary boundaries and reshape disciplines is itself fascinating for those seeking knowledge about the origins of particular intellectual communities. The stories told here are connected of course, but it is their differences that are most interesting to a historian of ideas. William Riker, one of the founders of modern analytical political science, writes of the ways in which game theory came into political thought in the 1950s. Similarly Vernon Smith, one of the originators of the field of experimental economics, writes of the development of that field and the particular direction the discipline took as a result of the then new theory of games. The history is presented at length as a case study of the development of a field in which game theory had some real impact on the underlying problematic and on the subsequent research agenda. The final paper, by the historian of science Robin Rider, treats the interconnection between game theory and operations research, where the latter field was likewise shaped and recast by the nature and limitations of the theory of games.

To close this brief prologue, I would like to acknowledge the help of Jeff Roggenbuck, whose index will be invaluable to readers of the volume, as well as that of Beth Eastlick, whose efforts on behalf of the *History of Political Economy* have made my work as editor of this volume much less a burden, much more a scholarly enterprise.

References

Aspray, William, and Philip Kitcher, eds. 1988. *History and Philosophy of Modern Mathematics*. Minneapolis: University of Minnesota Press.

Goodwin, Craufurd D. 1991. *Economics and National Security*. Durham, N.C.: Duke University Press.

Henderson, Lawrence J. 1935. *Pareto's General Sociology: A Physiologist's Interpretation*. Cambridge, Mass.: Harvard University Press.

Heyl, Barbara S. 1968. The Harvard Pareto Circle. *Journal of the History of the Behavioral Sciences* 4.4 (October): 316–34.

Ingrao, B., and G. Israel. 1990. *The Invisible Hand: Economic Equilibrium in the History of Science*. Cambridge, Mass.: MIT Press.

Kreps, David. 1990. *A Course in Microeconomic Theory*. Princeton: Princeton University Press.

Kruger, L., L. Daston, and M. Heidelberger, eds. 1987. *The Probabilistic Revolution*. Volume 1, *Ideas in History*. Cambridge, Mass.: MIT Press.

Kruger, L., G. Gigerenzer, and M. Morgan, eds. 1987. *The Probabilistic Revolution*. Volume 2, *Ideas in Science*. Cambridge, Mass.: MIT Press.

Leonard, Robert. 1991. War as a "Simple Economic Problem": The Rise of an Economics of Defense. In Goodwin 1991.

Mehrtens, Herbert, Henk Bos, and Ivo Schneider, eds. 1981. *Social History of Nineteenth-Century Mathematics*. Boston: Birkhauser.

Mirowski, Philip. 1989. *More Heat Than Light*. New York: Cambridge University Press.

———. 1991. When Games Grow Deadly Serious: The Military Influence on the Evolution of Game Theory. In Goodwin.

Morgan, Mary S. 1990. *The History of Econometric Ideas*. Cambridge: Cambridge University Press.

Punzo, L. 1991. The School of Mathematical Formalism and the Viennese Circle of Mathematical Economists. *Journal of the History of Economic Thought* 13.1:1–18.

Russett, Cynthia Eagle. 1966. *The Concept of Equilibrium in American Social Thought*. New Haven: Yale University Press.

Samuelson, Paul A. 1947. *Foundations of Economic Analysis*. Cambridge, Mass.: Harvard University Press.

Shackle, G. L. S. 1967. *The Years of High Theory*. Cambridge: Cambridge University Press.

Simon, Herbert. 1959. Review of "Elements of Mathematical Biology." *Econometrica* 27.3 (July): 493–95.

von Neumann, J. 1936. Über ein ökonomisches Gleichungssystem und eine Verallgemeinerung des Brouwerschen Fixpunksatzes in *Ergebnisse eines Mathematischen Kolloquiums, 1935–36*, edited by K. Menger. Leipzig and Vienna: Franz Deuticke. (Translation by G. Morton reprinted in 1945–46. A Model of General Economic Equilibrium. *Review of Economic Studies* 13.1:1–9.)

von Neumann, J., and O. Morgenstern. 1944. *The Theory of Games and Economic Behavior*. Princeton: Princeton University Press.

Weintraub, E. R. 1985. *General Equilibrium Analysis: Studies in Appraisal*. New York: Cambridge University Press.

———. 1991. *Stabilizing Dynamics: Constructing Economic Knowledge*. New York: Cambridge University Press.

Part 1 Creating Game Theory

The Early History of the Theory of Strategic Games from Waldegrave to Borel

Robert W. Dimand and Mary Ann Dimand

Probabilists have studied games of chance since the beginning of probability theory. Strategic games, whose outcome depends on the skill of the participants in choosing a strategy of play, received widespread attention among mathematicians and economists only with the publication in 1944 of the first edition of John von Neumann and Oskar Morgenstern's *Theory of Games and Economic Behavior*. The book marked an important advance, but it built upon an existing literature on strategic games to which both its authors had contributed. This paper examines the pre-1944 literature on the minimax theorem which holds that two-person zero-sum games with a finite number of pure strategies (or a continuum of pure strategies and continuous convex payoffs) are determined. Robert J. Aumann (1989) stresses the importance of the minimax solution of such games as a "vital cornerstone" for the development of game theory, noting that "the most fundamental concepts of the general theory—extensive form, pure strategies, strategic form, randomization, utility theory—were spawned in connection with the minimax theorem" (6–7) and that the Cournot-Nash concept of strategic equilibrium in non-cooperative *n*-person game theory is an outgrowth of minimax. We focus on the contributions made by the eminent French probabilist Émile Borel (1871–1956) in a series of papers from 1921 to 1927, placing his work on minimax solution of games of strategy in the context of earlier work as well as later papers by von Neumann (1928) and Jean Ville (1938).

The earliest minimax solution of a game was proposed more than two centuries before Borel's 1921 paper. Pierre-Rémond de Montmort wrote to Nicolas Bernoulli on 13 November 1713, in a letter translated in Baumol and Goldfeld (1968, 7–9), about a solution to a two-person version of the card game *le Her* proposed by James Waldegrave, then Baron Waldegrave of Chewton and later the first Earl Waldegrave (1684–1741). In this game,

> Peter holds a common pack of cards; he gives a card at random to Paul and takes one himself; the main object is for each to obtain a higher card than his adversary. The order of value is *ace, two, three, . . . ten, Knave, Queen, King*.
>
> Now if Paul is not content with his card he may compel Peter to change with him; but if Peter has a *King* he is allowed to retain it. If Peter is not content with the card which he at first obtained, or which he has been compelled to receive from Paul, he is allowed to change it for another taken out of the deck at random; but if the card he then draws is a *King* he is not allowed to have it, and must retain the card with which he was dissatisfied. If Paul and Peter finally have cards of the same value Paul is considered to lose. (Todhunter 1865, 106)

Paul would change any card lower than seven and hold any card higher than seven, while Peter would change any card lower than eight and hold any card higher than eight. If Paul always changed a seven, Peter would gain by adopting a rule of changing eight. However, if Peter always changed an eight, Paul would gain by always holding a seven instead of changing it. That is, in the doubtful cases (eight for Peter, seven for Paul), Peter would wish to follow the same rule (always hold or always change) as Paul, while Paul would wish to follow a rule opposite to that of Peter. Bernoulli thought that both players should change cards in the doubtful cases, while de Montmort concluded that no rule could be established.

Waldegrave considered the problem one of choosing a strategy that maximizes a player's probability of winning, whatever strategy may be chosen by his opponent. The resulting matrix of probabilities for a win by Peter, given the mixed strategy chosen, persuaded Waldegrave that a player could select a strategy assuring him of a certain outcome, while the other player could prevent him from doing better. Waldegrave concluded that Peter should hold cards of 8 and over (and change lower cards) with probability 5/8, and should change cards of 8 and under, holding higher cards, with probability 3/8. Paul should hold 7 and over with probability 3/8, and change 7 and under with probability 5/8 (see Kuhn 1968, 4–6; de Montmort [1713] 1968, 7–9; Todhunter 1865, 106–10). Norfleet W. Rives, Jr. (1975, 554) states incorrectly that Waldegrave derived a *matrix* of probabilities of winning from optimal *mixed* strategies. Rives also indicates that Waldegrave's solution involved the dealer always holding on 7 and over and the other player always changing on 8 and under, but this would mean that each player followed a pure rather than a mixed strategy and would result in a scalar probability of winning. Waldegrave's solution of *le Her* was a minimax solution, but he made no extension of his result to other games, and expressed concern that a mixed strategy "does not seem to be in the usual rules of play" of games of chance. He abandoned mathematics for diplomacy after leaving France for England in 1721. Although his mother was the natural daughter of the last Stuart king, James II, by Arabella Churchill (the sister of the first duke of Marlborough), Waldegrave served the House of Hanover as British ambassador to Vienna and to Versailles.

De Montmort published his correspondence with Jean and Nicolas Bernoulli, including his letter about *le Her,* as an appendix to the second edition of his *Essai d'Analyse sur les Jeux d'Hasard* (1713, 283–414; cf. Todhunter 1865, 113–34). This appendix became renowned for a letter from Nicolas Bernoulli to de Montmort stating the St. Petersburg paradox (see Daniel Bernoulli 1738). Despite this, Waldegrave's minimax solution of *le Her* remained largely unnoticed. The notions of a minimax solution, first proposed by Waldegrave, and of maximization of expected utility, diminishing marginal utility, and risk aversion put forward by Daniel Bernoulli (1738) in his analysis of the St. Petersburg paradox, lie at the core of game theory. These elements were not assembled until von Neumann and Morgenstern (1944) noted in an appendix on axiomatic utility theory in their second edition (1947, 629) that Bernoulli's moral expectation, which equated utility to the logarithm of wealth, satisfied their conditions for rationality. The two elements were not combined by Émile Borel, the rediscoverer of minimax solutions, even though he wrote extensively on the St. Petersburg paradox (see Jorland 1987, 177–78, 189 for references). Waldegrave's players attempt to maximize their probability of winning rather than either their expected monetary gain on each play or their "moral expectation" (expected utility) of such gains. In the game of *le Her* all of these are equivalent, so examination of this game would not have led Waldegrave to formulate a different objective function.

One writer did notice Waldegrave's solution. Isaac Todhunter reported it in his compendious *History of the Mathematical Theory of Probability* (1865, 106–10), although he also reported, without taking sides, the views of those who "asserted that it was impossible to say on which rule Paul should *uniformly act*" (429). Todhunter's book was recognized as the standard authority on its subject for nearly a century, so it might be expected that his discussion of Waldegrave's solution would attract notice, or that Todhunter's lengthy presentation of the work of Condorcet, Borda, and Laplace on the mathematical theory of elections (1865, 351–92, 432–34, 546–48, 618) would have been an early stimulus to social choice theory. Unfortunately, Todhunter's book, in the words of an admirer, M. G. Kendall, "is just about as dull as any book on probability could be. [Todhunter was] so unlike the colourful authors of whom he wrote, so meticulous in his attention to detail and so blind to the broad currents of his subject" (1963, 204–5) that his book was widely known rather than widely read. Literary style can make a difference to the development of a subject. If Todhunter had been a livelier writer, probabilists might have been thinking about minimax solutions of games of strategy and about voting theory in the late nineteenth century.

Waldegrave's solution of *le Her* was replicated, without mention of Waldegrave, by R. A. Fisher in "Randomisation, and an Old Enigma of Card Play" (1934). Although Harold Kuhn (1968) states that Fisher was "unaware of Waldegrave's work," Fisher discussed the views of de Montmort and Nicolas Bernoulli on *le Her,* dismissing de Montmort's conclusion as "unsatisfactory to common sense" (4n). Fisher's second sentence (1934, 294) refers to precisely those pages of Todhunter (1865, 106–10) that presented Waldegrave's solution.

While Émile Borel did not know of Waldegrave, he did have two results about strategic games to build on when he began publishing on the topic in the proceedings of the French Academy of Sciences in 1921, the year he was elected to the academy. First, Ernst Zermelo (1913) had proved that the game of chess is determinate, a result that has since been found to hold for other two-person zero-sum games of perfect information such as checkers, chinese checkers, and go, but not for such games as bridge or poker in which players have private information (Aumann 1989, 4). It is not, however, known what the optimal strategy is for a chess player, nor what the result would be if the optimal strategies were followed, an ignorance for which chess players must be grateful.

And second, consider Joseph Bertrand, who is best known to economists for his 1883 review of Cournot (1838) and Walras, often cited as the first review of Cournot (but see Dimand 1988 for an 1857 review by the Canadian mathematician J. B. Cherriman). Bertrand analyzed the game of baccarat in 1899, the year before his death. He considered whether a punter should draw another card when holding a count of five. He noted that the problem was psychological as well as mathematical, since the punter's decision on whether to draw for five depended on whether the banker expected him to do so. Bertrand examined, however, only the strategies of always drawing for five or never drawing for five, not a strategy of drawing for five on some fraction of plays (Bertrand 1924, cited in Borel 1924, 101).

Borel published four notes on strategic games and an erratum to one of them between 1921 and 1927. Three of these notes were translated by Leonard J. Savage in *Econometrica* (1953). Rives (1975, 559n) follows Maurice Fréchet (1953, 95n) in crediting Borel with seven notes on game theory in this period. However, the sixth and seventh items by Borel in Fréchet's bibliography (1953, 126) are the 1938 volume by Borel and collaborators in which Jean Ville (1938) appeared and a chapter by Borel in that volume, while one of the two untranslated notes is actually a 1923 article with the same title as the chapter that Savage translated from the 1924 volume by Borel. Borel was already an eminent probabilist when he turned his attention to strategic games, having stated the strong law of large numbers in 1909. He also had a prominent public career which brought him the *Croix de Guerre* in the First World War, the Medal of the Resistance in the Second, the portfolio of Minister of the Navy in 1925, and the Grand Cross of the Legion of Honor. His scientific honors included the presidency of the Institut de France and in 1955 the first gold medal of the Centre National de la Recherche Scientifique (Fréchet 1965, Knobloch 1987, Kramer 1981, 248–49).

Borel (1921) considered a two-person game of chance and strategy that was symmetric in the sense that if the two players adopted the same strategy, their chances of winning would be equal. The number of possible pure strategies was assumed to be finite. If A chooses method of play (pure strategy) C_i and B chooses method C_k, the probability of A winning is $a = 1/2 + \alpha_{ik}$ and the probability of B winning, $b = 1 - a$, is $b = 1/2 + \alpha_{ki}$, where $\alpha_{ik} + \alpha_{ki} = 0$, $\alpha_{ii} = 0$, and α_{ik} and α_{ki} are numbers between $-1/2$ and $+1/2$. Borel assumed that each player maximizes his probability of winning. As long as the payoffs are zero-sum and symmetric, this is equivalent to maximizing expected gains. Borel eliminated as "bad" those strategies C_i for which α_{ih} is negative or zero for every strategy C_h not already excluded as bad. Borel's criterion for the elimination of a bad strategy was, in the case of symmetry, as assumed in Borel 1921, equivalent to the criterion in current game theory for elimination of a weak or dominated strategy. A weak strategy is one that has a payoff less than or equal to that of some other strategy, no matter what the other player does. Borel therefore both anticipated the later criterion of elimination of weak strategies and noted that with such elimination, new pure strategies are likely to become weak.

If a strategy C_h existed such that α_{hk} was positive or zero for all k, that would be the best strategy. Borel noted that a best pure strategy may not exist, in which case it would be advantageous to adopt a mixed strategy, varying one's play across the n pure strategies remaining after the elimination of bad strategies. At any moment, the probability of player A playing strategy C_k is p_k, and the probability of player B playing strategy C_k is q_k. The probability of A winning is $a = 1/2 + \alpha$, where α is the summation across all i and k of $\alpha_{ik} p_i q_k$. For the case of three strategies, Borel noted that it is possible to select positive values for the p's such that α is zero, whatever q's may be chosen by the other player, and that the other player faces a symmetric problem of choosing q's.

Borel's 1921 solution of the choice of a mixed strategy when only three pure strategies are left after elimination of bad strategies is a minimax solution. However, Borel (1921) went on, "But it is easy to see that, once n exceeds 7, this circumstance [existence of a minimax solution] will occur only for particular values of the α_{ik}. In general, whatever the p's may be, it will be possible to choose the q's . . . in such a manner that α has any sign determined in advance" (98). Borel argued that, in the case of n greater than 7, as soon as one player has chosen a mixed strategy and his choice has been observed, the other player "may vary his play in such a manner as to have an advantage" (99). Borel presented no explanation or argument for this incorrect conjecture about the limitation of the minimax solution beyond the assertion that the result was easy to see.

Borel added, "It is easy to extend the preceding considerations to the case where the manners of playing form an infinite continuum" (99), but did not explain how extension from three strategies to an infinite number of pure strategies would be easy when extension to finite numbers greater than seven was not. He observed, "The problems of probability and analysis that one might raise concerning the art of war or of economic and financial speculation are not without analogy to the problems concerning games, but they generally have a much higher degree of complexity" (100). This complexity would include asymmetry and more than two players as well as more than seven possible pure strategies. In keeping with his belief that a player who knows his opponent's mixed strategy has an advantage when there are a finite number of strategies greater than seven, Borel concluded, "The only advice the mathematician could give, in the absence of all psychological information, to a player A whose adversary B seeks to utilize the preceding remarks is that he should so vary his plans that the probabilities attributed by an outside observer to his different manners of playing shall never be defined. . . . It seems that, to follow [this advice] to the letter, a complete incoherence of mind would be needed, combined, of course, with the intelligence necessary to eliminate those methods we have called bad" (100).

Borel began his 1924 paper "On Games That Involve Chance and the Skill of the Players" by noting the origin of the calculus of probabilities in the study of the simplest games of chance and suggested that the study of games involving the skill of players (choice of strategies) as well as chance should also begin with the simplest cases. He discussed Bertrand's analysis of baccarat, indicating the complexity of even baccarat when compared to a very simple, two-player symmetric game such as "paper, scissors, and stone" ("paper covers the stone, scissors cut the paper, stone grinds the scissors"), which Borel had chosen to study. Borel did not define the concept of symmetry he used in this paper, merely remarking that "neither player has a privileged position relative to the other. Further, it is evident, by reason of the same symmetry, that it is not possible to formulate advice permitting one of the players to win for sure; because if his adversary followed this same advice, he must also win for sure" (1924, 102).

In this game, "each player chooses at will, in an independent and secret manner, one of the three letters A, B, C. If the two players have chosen the same letter, the game is tied; otherwise it is agreed that A beats B, B beats C, and C beats A." Borel solved the game by maximizing the expected return for each player. If J chooses A, B, and C with probabilities x, y, and z, respectively, $x + y + z = 1$, and J′ chooses A, B, and C with probabilities x', y', and z', $x' + y' + z' = 1$, then J has an expected return $E = x(y' - z') + y(z' - x') + z(x' - y')$ and J′ has an expected return of $-E$. For any x, y, z, J′ can be sure of not losing systematically by adopting $x' = y' = z' = 1/3$, which sets the value of the game $E = -E = 0$, and similarly J can avoid systematic loss by choosing $x = y = z = 1/3$. Borel then solved the game for the case where J is the proprietor of a gambling house with an advantage over the punter J′ in that, when J has A and J′ has B, J receives the sum s, which is greater than 1. Borel extended his analysis to an asymmetric game in this instance and gave a good intuitive explanation of the house playing A less often than B or C: if the banker were to choose $x = y = z = 1/3$, "it would suffice for the punter never to play B in order that the advantage of the banker should be zero" (1924, 106). Borel also solved a symmetric two-person game with five possible pure strategies, showing that G, the mathematical expectations of the first player's winnings, would be zero, where

$$G = \sum \sum x_i y_k \alpha_{ik} = \sum x_i Y_i,$$

with x_i being the probability that the first player plays A_i and y_i the probability that the second player plays A_i.

However, Borel went on to state that "I will assume the hypothesis that, for n sufficiently larger, it is possible to so choose the constants that it is not possible to find positive nonzero values of the y's in such fashion that all the Y's should be positive or zero (or else all negative and zero). Under these conditions, whatever the y's may be, once they are determined, the x's can be chosen in such fashion that G will be positive" (114). As in his previous paper, Borel held, without offering any justification, that the minimax solution would not hold for large numbers of possible pure strategies, although he no longer asserted that the breakdown of the solution would occur as soon as n exceeded seven. This surprising claim does not follow from the analysis in his paper. It led Borel to conclude incorrectly that "the player who does not observe the psychology of his partner and does not modify his manner of playing must necessarily lose against an adversary whose mind is sufficiently flexible to vary his play while taking account of that of the adversary" (115). His analysis of symmetric games with $n = 3$ and $n = 5$ had shown, on the contrary, that in those games the first player has a mixed strategy to achieve $G = 0$ regardless of what his opponent does, and his opponent has a mixed strategy to prevent him achieving $G > 0$ no matter what the first player does.

Borel 1927 is very brief. Borel again defined the concepts of the symmetric game and the "tactic," or strategy. His previous definition of optimal play was a method that "gives the player who adopts it a superiority over every player who does not adopt it" (1921, 97). He now states that the central question of game theory is whether B can choose a tactic "so that player A if he knew that tactic, would nonetheless be unable to adopt a tactic of his own making G positive" (1927, 117). Borel's 1921 quest for a strategy that gives "superiority," which happened to coincide with a minimax strategy for the game he chose to examine, was replaced in 1927 by a search for a minimax strategy.

Borel concluded by indicating that he now believed that such a strategy exists for games that have as many as seven pure strategies and that it might exist more generally.

> The problem that arises is thus the following: *Determine the* α_{ik} *such that for all nonnegative* y_i *there are* Y_i *nonzero and not all of the same sign*. In this case player A can, by suitably choosing the x_i, give G the sign he wishes; i.e. win for certain on the average. This problem, unsolvable for $n = 3$ and $n = 5$, seems to me unsolvable also for $n = 7$. It would be interesting either to demonstrate that it is unsolvable in general or to give a particular solution. (Borel 1927, 117)

As von Neumann 1953 states, Borel has certainly not proved the minimax theorem. He may not even have stated it. Part of the minimax theorem is the statement that

$$\max_x \min_y G = \min_y \max_x G.$$

It is not clear that such equality occurred to Borel as being necessary for equilibrium, although his definition of optimality yields equality in the case of the class of games he examines. In a symmetric zero-sum game, Borel's equilibrium gives B the ability to prevent A from making G positive and A the ability to prevent B from making G negative. This automatically implies that equilibrium strategies by both parties set their expected payoff to zero. Borel had, however, correctly solved a nonsymmetric game for minimax strategy in Borel 1924.

Hugo Steinhaus, then a professor at the University of Lvov (then in Poland, now in Ukraine), contributed a short paper, "Definitions for a Theory of Games and Pursuit" to the December 1925 inaugural issue of a Lvov student journal that expired with its second issue. Steinhaus discussed a payoff function with strategies (which he called modes of play) as independent variables and defined the best mode of play as that which maximized the minimum expected payoff, including cases with a continuum of choices. He did not distinguish between pure and mixed strategies, as Borel had done. As Steinhaus recalled in a 1959 letter approving publication of an English translation of the paper, "After having found the concepts of minimax and maximin I was well aware that the minimax time of the pursuer is longer or equal to the maximin time of the pursued, but I did not know whether they are equal in all similar games" (1925, 108).

The first proof of the minimax theorem for two-person games of chance and skill with any finite number of pure strategies was given by John von Neumann (1928) in a paper presented to the Göttingen Mathematical Society on 7 December 1926. Von Neumann's proof was a complicated one, combining elementary and topological concepts in a manner not easy for the reader to follow, but it was a valid proof. In a footnote, von Neumann (25n) remarked that "while this paper was put into its final form, I learned of the note of E. Borel in the *Comptes Rendus* of Jan. 10, 1927. Borel formulates the question of bilinear forms for a symmetric two-person game and states that no examples for MaxMin < MinMax are known. Our result above answers his question." As Robert Leonard relates elsewhere in this volume, von Neumann sent Borel a note on his result, which Borel presented to the Académie des Sciences in June 1928.

In his 1937 paper, von Neumann gave an entirely topological proof of the existence of general competitive equilibrium, using Brouwer's fixed-point theorem to provide a much clearer and more elegant proof than his 1928 proof of the minimax theorem. The first elementary (nontopological) proof of the minimax theorem, using convexity arguments and the concept of a supporting hyperplane, is given by Jean Ville in a 1938 contribution to Émile Borel's *Traité du calcul des probabilités et de ses applications*. Ville (1938) also presented the first proof of the minimax theorem for the case of a continuum of possible pure strategies. The proof of the minimax theorem in von Neumann and Morgenstern 1944 is a nontopological one, based on the proof in Ville 1938 rather than on the one in von Neumann 1928. Hermann Weyl (1950) gave a simpler elementary proof of the minimax theorem, based on earlier work on convex polyhedra in a paper (Weyl 1935) that he had given at Göttingen in 1933 just before leaving to become a colleague of von Neumann at the Institute for Advanced Study in Princeton.

The subject of priority in scientific discovery is a prickly one, since original scientific contributions are prized and multiple independent discoveries do occur. Paul Samuelson (1989) refers to the instance of von Neumann's proof of the weak ergodic theorem in the early 1930s, shortly before George Birkhoff proved the strong ergodic theorem: "According to reliable legend, Birkhoff pulled strings so that von Neumann did not even get to publish his result before it would be eclipsed by the greater one" (120–21). Although Borel 1927 is mentioned in a footnote in von Neumann 1928, none of Borel's papers on strategic games was cited in von Neumann and Morgenstern 1944, in which Borel's name appears only in footnotes referring to the 1938 volume in which Jean Ville's paper appeared (von Neumann and Morgenstern 1947, 154n, 186n, 219n). Von Neumann (1953) responded to Fréchet's commentary on the Borel notes by stressing that his 1928 paper gave the first proof of the minimax theorem and that Borel had believed the theorem to be false when the number of possible strategies is large.

John von Neumann deserves the credit for the first general proof of the minimax theorem. James Waldegrave provided the first minimax mixed strategy solution of a two-person game of strategy more than two centuries before Borel and von Neumann, but his contribution was isolated and ignored. Waldegrave aside, Émile Borel gave the first modern formulation of a mixed strategy in 1921, Bertrand and Zermelo having dealt with pure strategies only. Borel found the minimax solution for two-person games of chance and skill with three or five possible strategies. While he initially held that games with more possible strategies would not have minimax solutions, by 1927 he considered this to be an open question, as he had not found a counterexample. Jean Ville, writing as a collaborator of Borel in 1938, provided the first elementary proof of the minimax theorem and extended the theorem to cases of infinitely many (continuous) strategies. Borel has received relatively little attention in the literature of game theory, for instance, only a single sentence in Aumann 1989 (6). His achievement in formulating mixed strategies, eliminating bad strategies, and finding minimax solutions of particular cases was a substantial contribution.

References

Aumann, R. J. 1989. Game Theory. In *The New Palgrave Game Theory*. Edited by J. Eatwell, M. Milgate, and P. Newman. New York: Norton.

Baumol, W. J., and S. Goldfeld. 1968. *Precursors in Mathematical Economics: An Anthology*, reprinted in *Scarce Works in Political Economy* No. 19. London: London School of Economics.

Bernoulli, D. [1738] 1954. Speciman theorie novae de mensura sortis. *Commentarii Academiae Scientiarum Imperialis Petropolitanae* 5:175–92. Translated by L. Somer as Exposition of a New Theory on the Measurement of Risk. *Econometrica* 22 (January): 23–36.

Bertrand, Joseph. 1924. *Calcul des probabilités, éléments de la théorie des probabilités*, 3d ed. Paris: Gauthier-Villars.

Borel, É. [1921] 1953. La théorie du jeu et les équations, intégrales à noyau symétrique gauche. *Comptes Rendus de l'Académie des Sciences* 173 (December): 1304–8. Translated by L. J. Savage as The Theory of Play and Integral Equations with Skew Symmetric Kernels. *Econometrica* 21 (January): 97–100.

———. [1924] 1953. Sur les jeux où interviennent l'hasard et l'habilité des joueurs. In *Théorie des probabilités*. Paris: Librairie Scientifique, J. Hermann. Translated by L. J. Savage as On Games that Involve Chance and the Skill of Players. *Econometrica* 21 (January): 101–15.

———. [1927] 1953. Sur les systèmes de formes linéaires à déterminant symétrique gauche et la théorie générale du jeu. *Comptes Rendus de l'Académie des Sciences* 184:52–53. Translated by L. J. Savage as On Systems of Linear Forms of Skew Symmetric Determinant and the General Theory of Play. *Econometrica* 21 (January): 116–17.

Borel, É., et al. 1925–39. *Traité du calcul des probabilités et de ses applications*, 4 vols., 18 fascicles. Paris: Gauthier-Villars.

Cournot, A. A. [1838] 1927. *Recherches sur les principes mathématiques de la théorie des richesses*. Paris: Hachette. Translated by N. T. Bacon as *The Mathematical Principles of the Theory of Wealth*. New York: Macmillan.

de Montmort, P. R. 1713. *Essai d'analyse sur les jeux de hasard*, 2d ed. Paris: Quilau.

———. [1713] 1968. [Excerpt from de Montmort 1713]. In *Precursors in Mathematical Economics: An Anthology*, edited by W. J. Baumol and S. Goldfeld. Reprinted in *Scarce Works in Political Economy* No. 19. London: London School of Economics.

Dimand, R. W. 1988. An Early Canadian Contribution to Mathematical Economics: J. B. Cherriman's 1857 Review of Cournot. *Canadian Journal of Economics* 21.3 (August): 610–16.

Fisher, R. A. 1934. Randomisation, and an Old Enigma of Card Play. *Mathematical Gazette* 18:294–97.

Fréchet, M. 1953. Émile Borel. *Econometrica* 21 (January): 95–99, 118–24.

———. [1965] 1972. La vie et l'oeuvre d'Émile Borel. *Monographies de l'Enseignement mathématique* 14. Reprinted in *Oeuvres de Émile Borel*, I. Paris: Centre National de la Recherche Scientifique.

Jorland, G. 1987. The St. Petersburg Paradox 1713–1937. In *The Probabilistic Revolution*, edited by L. Kruger, L. Daston, and M. Heidelberger. Cambridge, Mass.: MIT Press.

Kendall, M. G. 1963. Isaac Todhunter's *History of the Mathematical Theory of Probability*. *Biometrika* 50:204–5.

Knobloch, E. 1987. Émile Borel as a Probabilist. In *The Probabilistic Revolution* I, edited by L. Kruger, L. Daston, and M. Heidelberger. Cambridge, Mass.: MIT Press.

Kramer, E. 1981. *The Nature and Growth of Modern Mathematics*. Princeton: Princeton University Press.

Kuhn, Harold. 1968. Introduction to de Montmort. In *Precursors in Mathematical*

Economics: An Anthology, edited by W. J. Baumol and S. Goldfeld. Reprinted in *Scarce Works in Political Economy* No. 19. London: London School of Economics.

Rives, Jr., N. W. 1975. On the History of the Mathematical Theory of Games. *HOPE* 7.4 (Winter): 549–65.

Samuelson, P. A. 1989. A Revisionist View of von Neumann's Growth Model. In *John von Neumann and Modern Economics*, edited by M. Dore, S. Chakravarty, and R. Goodwin. Oxford: Clarendon.

Steinhaus, H. [1925] 1960. Definitions for a Theory of Games and Pursuit (in Polish). *Mysl Akademicka Lwow* 1:13–14. Translated by E. Rzymovski with an introduction by H. Kuhn, in *Naval Research Logistics Quarterly* 7.2 (June): 105–8.

Todhunter, I. [1865] 1965. *A History of the Mathematical Theory of Probability from the Time of Pascal to that of Laplace*. Cambridge: Cambridge University Press. Reprinted, Bronx: Chelsea.

Ville, J. 1938. Sur la théorie générale des jeux où intervient l'habilité des joueurs. In *Applications des jeux de hasard*, by É. Borel et al. 4.2:105–13.

von Neumann, J. [1928] 1959. Zur theorie der gesellschaftsspiele. *Mathematische Annalen* 100:295–320. Translated by S. Bargmann as On the Theory of Games of Strategy. In *Contributions to the Theory of Games* 4, edited by A. W. Tucker and R. D. Luce, *Annals of Mathematical Studies* 40. Princeton: Princeton University Press.

———. [1937] 1945. Über ein Oekonomisches Gleichungssystem und eine Verallgemeinerung des Brouwerschen Fixpunktsatzes. *Ergebnisse eines Mathematischen Seminars*, edited by K. Menger, Vienna. Translated by G. Morton as A Model of General Economic Equilibrium. *Review of Economic Studies* 13.1:1–9.

———. 1953. Communications on the Borel Notes. *Econometrica* 21 (January): 124–25.

von Neumann, J., and O. Morgenstern. 1944. *The Theory of Games and Economic Behavior*. Princeton: Princeton University Press.

———. 1947. *The Theory of Games and Economic Behavior*, 2d ed. Princeton: Princeton University Press.

Weyl, H. [1935] 1950. Elementare Theorie der konvexen Polyeder. *Commentarii Mathematici Helvetici* 7:290–306. Translated by H. W. Kuhn as The Elementary Theory of Convex Polyhedra. In *Contributions to the Theory of Games* 1, edited by H. W. Kuhn and A. W. Tucker, *Annals of Mathematical Studies* 24. Princeton: Princeton University Press.

———. 1950. Elementary Proof of a Minimax Theorum Due to von Neumann. *Contributions to the Theory of Games* 1, edited by H. W. Kuhn and A. W. Tucker, *Annals of Mathematical Studies* 24. Princeton: Princeton University Press.

Zermelo, Ernst. 1913. Über eine Anwendung der Mengenlehre auf die theorie des Schachspiels. *Proceedings, Fifth International Congress of Mathematicians* 2:501–4.

Creating a Context for Game Theory

Robert J. Leonard

Introduction

After decades of hesitation, the theory of games now plays a central role in economic theory. New textbooks on microeconomics no longer relegate Nash equilibrium to a section on "other topics," but place the formalization of interaction at the theory's very heart (see Kreps 1990). Nor has this been confined to the theoretical core: the "new" industrial organization, for example, differs from the older "structure-conduct-performance" paradigm in its emphasis on the game-theoretic aspect of firms' decisions (see Jacquemin 1987, Tirole 1988). Regardless of how one views the worth of these developments—and in this regard there is considerable debate (see Fisher 1989)—the ideas of game theory have permeated economics in a circuitous manner. While the stylized historical précis locates the seminal ideas with von Neumann and Morgenstern (1944) and their purification and refinement by Nash, Shapley, Aumann, and others in the intervening period, all cumulatively contributing to the position in which we find ourselves today, the adoption of the game paradigm by economic theorists has not been a smooth process. Indeed the "new" ideas of 1944 were primarily appropriated and developed not by economists but by mathematicians. The latter found theoretical games interesting because of their links with existing fields in mathematics and statistics and because the new ideas constituted a particularly rich bed of mathematical ore: the opportunity for constructing new theo-

For helpful commentary and discussion on this material my thanks go to Craufurd Goodwin, Jim Leitzel, Matthew O'Meagher, Urs Rellstab, and, particularly, Roy Weintraub. Financial support from the Pew Charitable Trusts facilitated the basic research. This included the consultation of the Morgenstern Papers at Duke University, the von Neumann Papers at the Library of Congress, and interviews with former RAND mathematicians, including Samuel Karlin and Edward Harris, to whom I am also grateful.

rems seemed great. They received support for such work because of the perception, in certain quarters, that game theory had potential military application. What is particularly interesting is that the mathematical work following von Neuman and Morgenstern 1944 concentrated not on those aspects of games given most attention in that book, but on earlier work by both von Neumann and some French mathematicians whom he seems to have ignored. Game theory's initial development rested, ironically, not on the material in the *Theory of Games*, but in the disparate papers that preceded it. In part 1 of this article, I discuss the disconnected contributions to mathematical games of the pre-1944 period, showing how, in the absence of any discursive mathematical community, game "theory" remains something of a misnomer. Part 2 examines the social, institutional, and mathematical transformations brought about by World War II, which created a context in which game theory became significant. After all the initial noise died down, these mathematicians took to exploring primarily pre-Morgenstern game theory, i.e., two-person noncooperative games. Furthermore, this postwar work was done with von Neumann's sanction and encouragement, even though he alone had developed the mathematics of cooperative games, had devoted the bulk of the *Theory of Games* to that topic, and saw the analysis of n-person games as the theory's crowning achievement.

The history of game theory has only recently begun to receive significant academic attention. Rives 1975 provides a useful introduction but does not explore any particular aspect in great detail. Mirowski 1991 treats some of the subject matter of this article, arguing that the military influence on games was not only significant, but retarding, in that it prevented the theory's proper or "logical" development. This article explores in considerable detail the seminal work by European mathematicians in the first half of this century, and then links it to the developments following the appearance of von Neumann and Morgenstern 1944. The latter work is shown to be significant not in that it *itself* became the object of direct mathematical attention, but in that it focused the attention of mathematicians and their patrons on an inchoate body of applied mathematics which World War II had made relevant.

1. Exploration: The Theory of Games to 1945

In 1953 there appeared in the columns of *Econometrica* (see Fréchet 1953) a restrained but firm debate on the legitimacy of John von Neumann's position as "initiator" of game theory. The challenge, posed by French mathematician Maurice Fréchet, implied that due credit had not been paid to his senior colleague Émile Borel. The latter, Fréchet pointed out, had written on game theory in the early 1920s prior to von Neumann, and while he had not proved the central minimax theorem, he had raised the question of its validity and had speculated on the ultimate application of such ideas to economic and military problems. Von Neumann wrote a stiff rejoinder claiming that until his 1928 proof of the theorem, "there was nothing worth publishing" (125). Fréchet remained committed to his claim, countering that Borel's early speculations provided "an open door," through which von Neumann could walk.

I take this debate as the point of entry into the history of game theory for the following reasons. First, it is a sign that something of theoretical significance had happened. There had emerged a set of ideas in which several parties had different, often conflicting, interests: there was something worth arguing about. Second, to the extent that it indicates what historical aspects the participants themselves found interesting, it helps us cast an interpretative net back over the period in question. Part 1 begins with an exploration of the period referred to by Fréchet. In particular, I examine the relevant work of Borel, von Neumann, and also the Polish mathematician Hugo Steinhaus. I also carry the inquiry through the 1930s, covering the passage of game theory to the United States, up to the publication of von Neumann and Morgenstern 1944. My concern, it must be emphasized, is not to reopen the priority debate. Rather, I illustrate with equal emphasis that, for the most part, there was negligible interaction between the individual mathematicians concerned. To the extent that a mathematical theory is given life by a discursive community arguing and contributing to the set of ideas in question, the mathematical analysis of games before World War II was a particularly lifeless affair. Not only was the number of interested people involved very small but, for all intents and purposes, they remained incommunicado. Only with the appearance of von Neumann and Morgenstern 1944 did this state of affairs begin to change. Throughout, to the greatest extent practicable, I describe the intellectual context of the ideas in question.

Borel's Wager

Like so many of the protagonists in our short historical excursion, Borel led a life that was, to put it mildly, fuller than that of many of his contemporaries (see Collingwood 1959, Fréchet 1965). His brilliance was recognized early, when at age eighteen, he won first prize in the Concours Général and achieved first place in the admission lists for both the École Normale and the École Polytechnique. For the rest of his career, he was prodigiously active in the spheres of both academics and politics.

Borel was directly influenced by mathematician Gaston Darboux in his decision to become a *Normalien* and pursue a career in research. His 1894 doctoral thesis on the theory of functions stemmed from a theme derived from Darboux and contained many of the seminal ideas that Borel would soon develop. During the period to 1905, he made several contributions of great significance, including the theory of measure, later developed by Lebesgue, the theory of divergent series, and an elementary proof of Picard's Theorem, which mathematicians had sought for over seventeen years. The culmination of this period was the beginning of a voluminous series on the theory of functions, edited and directed by Borel, to which he himself contributed five volumes. Under his directorship, fifty volumes of these "Borel Tracts" appeared.

His interests broadened during the first decade of this century to encompass probability theory, where he introduced the notion of enumerable probability[1] and his strong law of large numbers. In addition, with money he had received in academic awards, in 1906 he founded the *Revue du Mois*, a popular monthly, to which he contributed articles of scientific, philosophical, and sociological interest. He also undertook the editorship of a series of books intended to popularize scientific ideas, to which he himself contributed (with Painlevé) *l'Aviation* (1910) and *le Hazard* (1914). After World War I he entered public life, becoming a member of Parliament for twelve years, occasionally holding various positions of higher office, and all the while writing prolifically in mathematics. In 1921, having exchanged his chair of Theory of Functions at the École Normale for that of Theory of Probability and Mathematical Physics, formerly held by Poincaré, he began to edit and contribute to the monumental *Traité du Calcul des Probabilités et de ses Applications*, a series of monographs intended to "organize and expound the whole mathematical theory of probability and its applications as it

1. The issue of enumerable probability is concerned with the probability of events that depend upon the totality of an infinite sequence of random variables. While Bernoulli and others had examined this problem using the asymptotic properties of the probabilities, Borel was the first to consider the totality. His theorem showed that the probability for the totality depended on the convergence or divergence of the sum of probability sequences (see Collingwood 1959).

had developed up to that time" (Collingwood 1959, 488). This undertaking occupied the next fifteen years, till World War II, after which the seventy-four-year-old Borel began to write another fifty notes, papers, and books, mainly presenting mathematical and physical ideas to nonspecialist readers.

His first published work on the mathematics of games dates from 1921, by which time he had become absorbed in probability theory. In a series of notes written throughout the 1920s, Borel gives a reasonably systematic, if ultimately speculative, treatment of what would later be known as two-person symmetric zero-sum games. He clearly considered this work important enough for presentation to the august Académie des Sciences, to which he had been elected in 1921: three of the five notes appear in the academy's proceedings, the *Comptes Rendus*. As some of his observations are repeated from one paper to the next, I shall confine my discussion to the three most important (1921, 1924, 1927) in an attempt to capture both the letter and spirit of Borel's inquiry. To render this work tractable, I make a distinction between Borel's specific speculation on the possibility of a deeper mathematical theorem, which I examine first, and his general thoughts, by which is intended his illustration of concrete examples of games and his musing on the broader applicability of the general framework.

Borel 1921 defines certain concepts and creates the framework upon which the subsequent notes would build, albeit with frequent changes in notation.[2] First, Borel suggests that we consider a game "in which the winnings depend both on chance and the skill of the players," unlike such games as dice where skill does not influence the outcome. Defining a "method of play" as "a code that determines for every possible circumstance . . . what the person should do," Borel asks "whether it is possible to determine a method of play better than all others" (Fréchet 1953, 97).

Consider a game with two players A and B, who choose strategy ("method") C_i and C_k, respectively. Each have the same set of n strategies available. Given the strategies chosen, the entries in the matrix represent A's probability of winning the game: no winnings are transferred between players.[3] The numbers α_{ik} and α_{ki} are contained between

2. As far as possible, I use consistent notation. I also use matrices to a greater extent than Borel, in order to portray his ideas more easily.

3. In his later work, Borel uses payoffs in the matrix rather than probabilities, in the now conventional manner.

$1/2$ and $+1/2$, and satisfy $\alpha_{ik} + \alpha_{ki} = 0$. Also, $\alpha_{ii} = 0$. The game is symmetric and fair.

		Player B			
		C_1	C_2		C_n
	C_1	$1/2 + \alpha_{11}$	$1/2 + \alpha_{12}$	$\cdots$	$1/2 + \alpha_{1n}$
Player A	C_2	$1/2 + \alpha_{21}$	$1/2 + \alpha_{22}$	$\cdots$	$1/2 + \alpha_{2n}$
	$\cdots$	$\cdots$	$\cdots$	$\cdots$	$\cdots$
	C_n	$1/2 + \alpha_{n1}$	$1/2 + \alpha_{n2}$	$\cdots$	$1/2 + \alpha_{nn}$

Players are assumed to automatically cast aside "bad" strategies, for example, methods of play that guarantee a probability of winning of less than half.[4] Having done this, the question is how the remaining strategies might be employed in the best manner possible. Borel suggests that a player can act "in an advantageous manner by varying his play," i.e., play a mixed strategy. C_k is played with probability x_k by A and y_k by B, where

$$\sum_1^n x_k = 1 = \sum_1^n y_k .$$

Given this, A's expected probability of winning is

$$\sum_1^n \sum_1^n (1/2 + \alpha_{ik}) x_i y_k = 1/2 + \alpha$$

$$\text{where } \alpha = \sum_1^n \sum_1^n \alpha_{ik} x_i y_k$$

and B's probability of winning is thus $1/2 - \alpha$.

Borel now considers the case where $n = 3$, given by the following matrix:

		Player B		
		C_1	C_2	C_3
	C_1	$1/2 + \alpha_{11}$	$1/2 + \alpha_{12}$	$1/2 + \alpha_{13}$
Player A	C_2	$1/2 + \alpha_{21}$	$1/2 + \alpha_{22}$	$1/2 + \alpha_{23}$
	C_3	$1/2 + \alpha_{31}$	$1/2 + \alpha_{32}$	$1/2 + \alpha_{33}$

Since no strategy is "bad," it must be the case that α_{23}, α_{31}, and α_{12} are all the same sign. He asks, is there any way that A can choose x_i

4. Borel suggests that strategies offering a probability of only 1/2 should also be rejected, but it is difficult to see why a player would do this and then proceed to seek a guarantee of 1/2 with the remaining strategies.

in such a manner as to ensure $\alpha \geq 0$, for any vector y_k? Alternatively put, if A knows the probabilities that B is going to employ, can he or she find a mixed strategy that will ensure an even chance of winning? Such positive numbers, x_1, x_2, x_3, Borel claims, are always easy to find: in a two-person symmetric game with three strategies available to each player, each player can ensure an even chance of victory.

In 1924, Borel extends this analysis, slightly modified, to the case of $n = 5$, where each player can choose from five strategies. Here, the entries in the matrix are the payoffs to player A. The same assumptions hold:

$$\alpha_{ij} = -\alpha_{ji} \text{ and } \alpha_{ij} = 0, \text{ if } i = j.$$

		Player B				
		C_1	C_2	C_3	C_4	C_5
	C_1	0	α_{12}	α_{13}	α_{14}	α_{15}
	C_2	$-\alpha_{12}$	0	α_{23}	α_{24}	α_{25}
Player A	C_3	$-\alpha_{13}$	$-\alpha_{23}$	0	α_{34}	α_{35}
	C_4	$-\alpha_{14}$	$-\alpha_{24}$	$-\alpha_{34}$	0	α_{45}
	C_5	$-\alpha_{15}$	$-\alpha_{25}$	$-\alpha_{35}$	$-\alpha_{45}$	0

Now A's expectation becomes

$$\sum_1^5 \sum_1^5 x_i y_k \alpha_{ik} = \sum_1^5 x_i Y_i$$

where

$$Y_i = \sum_1^5 y_k \alpha_{ik}.$$

Taking, for analytical convenience, the skew symmetric matrix of the payoffs to be the following:

	Player B				
	0	1	a_1	$-a_4$	-1
	-1	0	1	a_2	$-a_5$
Player A	$-a_1$	-1	0	1	a_3
	a_4	$-a_2$	-1	0	1
	1	a_5	$-a_3$	-1	0

the Y_i, player B's expected payoff given that A plays strategy i, can be written:

$$Y_1 = y_2 + a_1y_3 - a_4y_4 - y_5$$

$$Y_2 = -y_1 + y_3 + a_2y_4 - a_5y_5$$

$$Y_3 = -a_1y_1 - y_2 + y_4 + a_3y_5$$

$$Y_4 = -a_4y_1 - a_2y_2 - y_2 + y_5$$

$$Y_5 = y_1 + a_5y_5 - a_3y_3 - y_5$$

Borel now asks, Is it possible for player B to choose the vector of y_i such that each of the Y_i is no less than zero? He then devotes the rest of the paper to showing that appropriate probabilities y_i can always be found. Depending on the values of the a_i in the payoff matrix, B can keep all the Y_i to zero, thus ensuring no advantage to A. Borel concludes that with $n = 5$, "nothing essentially new happens compared to the case where there are three manners of playing" ([1924] 1953, 114), i.e., each player can ensure an expected payoff of zero. But speculating as to whether this is likely to hold for n arbitrarily large, Borel is pessimistic and suggests that it will not always hold. However, three years later, in another note (Borel [1927] 1953) presented to the Académie, he reports an extension of his analysis giving rise to greater optimism: what has held for three and five strategies seems also to hold for seven, and it would thus "be interesting either to demonstrate that it is unsolvable in general or to give a particular solution" (117).[5]

While Borel may have concluded his search for a deeper theorem on a speculative note, his work of the 1920s is notable in certain other respects, especially in the provision of general concepts and examples. First, introducing the infinite game, where strategies are drawn from a continuum, he shows how the continuous analogue of player A's expected payoff is expressible as a Stieltjes integral:

$$\alpha = \int_{-\infty}^{\infty}\int_{-\infty}^{\infty} f(C_{\mathrm{A}}, C_{\mathrm{B}})d\phi_{\mathrm{A}}(C_{\mathrm{A}})d\phi_{\mathrm{B}}(C_{\mathrm{B}})$$

where $f(C_{\mathrm{A}}, C_{\mathrm{B}})$ is the function relating A's payoff to the strategies chosen, and $\phi_{\mathrm{A}}(C_{\mathrm{A}})$ and $\phi_{\mathrm{B}}(C_{\mathrm{B}})$ are A's and B's respective cumulative distribution functions over strategy space. The example offered of such a game is what later was to become known as a game on the unit square: in this case, each player chooses three real numbers summing to 1, the

5. Borel's analysis is confined to games with odd numbers of strategies because of his use of determinants of skew-symmetric matrices to calculate the optimal mixed strategy.

winner being the one with two choices of greater value than the opponent's. While Borel simply describes the game here, a decade or so later, in 1938, his student Jean Ville would extend the minimax theorem to such games.

Second, Borel looks at specific examples of finite games such as "paper, scissors, stone" ([1924] 1953, 102) and shows in detail how the calculation of the optimal mixed strategies depends on relative payoffs. For example, if the payoff to A for a particular strategy is relatively large, then the probability attached to it in the optimal mixed strategy will be correspondingly low: otherwise, B could gain by anticipating A's emphasis on the favored strategy.

Third, he considers the broader application of these ideas to the nonmathematical realm: "The problems of probability and analysis that one might raise concerning the art of war or of economic and financial speculation are not without analogy to the problems concerning games" ([1921] 1953, 100). He cautions restraint in this regard, however, saying that such matters are highly complex and that mathematical calculation can at best be a supplement to strategic cunning. As we shall see below, this trace of scepticism in Borel would grow even stronger with time.

In his final related communication to the Académie, in May 1928, Borel supplied the answer to the question that had occupied him in the notes discussed above, that is, concerning the existence of a "best" way to play. However, this clarification came not as a result of further work by Borel himself, but in the shape of a note from somebody who claimed to have proved the minimax theorem in Göttingen two years previously. This, of course, was John von Neumann, then in his mid-twenties and thirty years Borel's junior (see von Neumann 1928a). Before turning to von Neumann, however, I shall consider the related contemporaneous work of Hugo Steinhaus at Lvov in Poland. While Borel and von Neumann were at least aware of each other's work and common interest, Steinhaus, it appears, labored in total isolation.

A Pole Apart

The period between the two world wars was one during which mathematics flourished in Poland (see Kuratowski 1980). Under the tutelage of such figures as Zaremba and Sierpinski, there emerged from the universities in Lvov and Wroclaw a number of capable young mathematicians including Banach, Ulam, and Steinhaus. Born in 1887 in Jaslo, Poland,

Steinhaus spent one year studying at the university in Lvov and then took off to Göttingen, where he completed his doctorate in 1911 under David Hilbert. Following this, he lived as an independent scholar in Jaslo, until the outbreak of war, when he joined the Legions. In 1916 he joined the faculty at Lvov, became full professor by 1923, and stayed until again interrupted by war in 1941. In 1945 he moved to the scientific center at Wroclaw, where he remained until his death in 1972. In both Lvov and Wroclaw, Steinhaus was a leader among the groups engaged in mathematical research (see Kac 1985).

Steinhaus's mathematical interests were wide-ranging in both theoretical and applied areas. In the former, he was active in the theory of trigonometrical series, functional analysis, orthogonal series, the theory of real functions, and, perhaps most famously, in sequences of linear operations, for which he is remembered as a collaborator of Banach. In applied areas, he published on probability theory and on the application of mathematics to questions in medicine, electricity, biology, geology, and anthropology. It was, no doubt, this latter taste for the mathematics of applied problems that led Steinhaus to games.

The work in question is a single article, "Definitions for a Theory of Games and Pursuit," which appeared in 1925 in the first issue of "an ephemeral pamphlet called 'Mysl Akademicka' " (Steinhaus 1960, 108). A short-lived periodical edited by Lvov students, its first issue was also its second to the last! Here, Steinhaus claims that he is concerned with the construction of mathematical definitions for a group of problems that lie "beyond the strict boundaries of mathematics" (106). The problems discussed are chess, naval pursuit, and card-playing, and the thread that binds them together is their use by Steinhaus to motivate the notion of minimax play (without actually terming it such). He introduces the concept of a "mode of play" which connotes, for either player, "a list of all possible circumstances with a preferred move for each" (106). Considering chess, he modifies the game by placing a limit, known to both players, on the total number of moves permitted. Should White not win before this limit is reached, Black wins. Black's aim, therefore, is to adopt the strategy that prolongs his defense, while White's is to keep the length of play as short as possible by winning as quickly as possible. Given that Black chooses a strategy to maximize the duration, White chooses the strategy that keeps this maximum to a minimum. In exactly the same manner, Steinhaus discusses two ships in pursuit: one chasing, the other fleeing, both at a given speed. The pursuer's aim is to close the angle between its line of steering and line of sight, thereby minimizing the time in pursuit. The evader's aim is the opposite. Each ship's strategy is a function of both ships' coordinates. At any moment, given that the evader has chosen the time-maximizing strategy, the hunter will respond by attempting to minimize it. Finally, Steinhaus notes that in card games a similar tussle is involved as each player tries to reduce the expected gain of his or her opponent.

While conceding that only definitions have been provided, Steinhaus notes that these are essential for the next stage, the calculation of best play. However, actually finding the "best move," "best pursuit," or "best way of playing" involves "enormous difficulties." Pursuit, for example, would require us "to use the calculus of variations on a very difficult problem of mathematical analysis," while even the simplest card games "lead to very involved combinatorial calculations" (107). Had Steinhaus learned the following year that the mathematical consideration of games was the subject of discussion among the Göttingen group, centered on his doctoral supervisor Hilbert, his paper might not have been forgotten. As it was, it gathered dust for decades.[6]

New Boy at Göttingen

Born to a wealthy Jewish banking family in Budapest in 1903, "Jansci" Neumann was the eldest of three brothers.[7] Among these he was "the most aggressive one, the least sentimental, the most thoughtful," staying away from childish games, preferring to read, study, or calculate or, during World War I, play "elaborate battles with toy soldiers" (Heims 1980, 41). The young von Neumann was precocious and, shortly after entering the Lutheran Gymnasium at age ten, was recognized by his mathematics teacher as a child prodigy. At the teacher's suggestion, the

6. Steinhaus's paper was not translated into English until 1960, when it was published in the *Naval Research Logistics Quarterly*. The paper is introduced there by Harold Kuhn who explains that the Polish version was secured by Stan Ulam. An early reference to the paper appears, however, in a letter from Oskar Morgenstern to Olaf Helmer at RAND in October 1952, in which he states that a friend was writing to Steinhaus for a copy of the paper (see OMPD, Box 14, File RAND). In a letter accompanying the published paper, Steinhaus offers some background to his search for the paper. Not until 1957 did he retrieve the paper, and then only through a colleague who secured a copy of the journal in Lvov, which was by then part of the Soviet Union.

7. The family name Neumann became von Neumann when John's father, Max, was ennobled by Emperor Franz Joseph in 1913 "for his contributions to the economic development of Hungary" (see Aspray 1990, 254 n. 4; Heims 1980, 29–32).

child's father arranged additional tutoring from a mathematics lecturer at the University of Budapest. By the time he left secondary school, he had won the nationwide Eotvos prize for "excellence in mathematics and scientific reasoning" (44).

In 1921, at seventeen, von Neumann enrolled at the University of Budapest but promptly left the country for Berlin, thereafter returning to Budapest to take the necessary university exams or visit his family.[8] Spending 1921–23 at the University of Berlin, he was influenced by David Hilbert's former student Erhard Schmidt and by Albert Einstein from whom he took lectures on statistical mechanics. He kept company with fellow Hungarian émigrés including Eugene Wigner, Leo Szilard, and Dennis Gabor, all part of the "Hungarian Phenomenon," which was later to have an enormous impact on American physics and mathematics. During this time, von Neumann also made contact with Hilbert at Göttingen, the "mecca of German mathematics," beginning a collaboration which would last several years and influence von Neumann for life. From Berlin, he went to the Swiss Federal Institute of Technology at Zurich where he took a degree in chemical engineering in 1925. There he fell in with George Polya, another Hungarian mathematician who would later come to Stanford University, and Hermann Weyl, a German mathematician who was to be one of the first residents at Princeton's Institute for Advanced Study (IAS). In 1926, von Neumann received his doctorate in mathematics from Budapest and spent the rest of the year in Göttingen; the following year he was appointed Privatdozent at the University of Berlin. Throughout this entire period, he maintained close contact with Hilbert.

Given the importance of von Neumann to our story, it is worth portraying the historical developments in mathematics in which his mentors at Göttingen had crucial roles.[9] Over the nearly fifty years from 1895 during which he reigned at Göttingen, Hilbert made mathematical contributions that arguably, among mathematicians, exalted him to the

8. In Fermi (1968), Eugene Wigner, Hungarian contemporary of von Neumann and later atomic physicist in the United States, comments on the tendency for young Hungarians to leave early: "[While] Hungarian high schools were excellent, . . . the universities were very poor. . . . It was symbolic of the bureaucratic influence . . . that professors at universities kept their libraries locked up and a student who needed to consult a book was often obliged to borrow the key from his professor" (54).

9. For a fine account of this period, the reader is referred to Constance Reid's 1970 biography of Hilbert.

rank of an Archimedes, Newton, or Gauss. By 1902, he had made his mark on, *inter alia,* invariant theory, the calculus of variations, and the foundations of geometry. In the last, his axiomatization of Euclidean geometry (*The Foundations of Geometry*, 1899) signaled the beginning of a lifelong preoccupation with the way in which proofs in mathematics are related to the axioms on which they are based. True rigor, for Hilbert, required that axioms be complete, in the sense that all theorems be derivable from them; independent, in that the removal of any axiom would make it impossible to prove at least some of the theorems; and consistent, so that no contradictory theorems could be established using such axioms. In this, he was further galvanized by Zermelo's and Russell's independent observations of a fundamental antimony, or paradox, in set theory and called for the mathematical investigation of proof itself.[10] This became known as metamathematics or proof theory. Just as Hilbert was concerned with the logical rigor of mathematics, so too was he disturbed by the apparent lack of order in the constructions of the physicists, which at this time were growing rapidly. The turn of the century saw Hertz's proof of the existence of electromagnetic waves; Roentgen's discovery of X rays; the Curies' radioactivity; J. J. Thomson's electron; Einstein's special theory of relativity; and Max Planck's quantum theory. Hilbert "perceived the pressing necessity for investigation to determine whether these diverse principles were compatible with one another and in what relation they stood" (quoted in Reid 1970, 127). Thus, beginning in 1912, he turned his attention to the mathematics underlying purely physical phenomena, beginning with kinetic gas theory and then elementary radiation theory, in each case constructing, from the axiomatic base up, a mathematical theory consistent with the physics.

After World War I, Hilbert's attention reverted to the exploration of foundations in response to the view of mathematics known as intuitionism, then being propagated by Dutch mathematician L. Brouwer. Briefly stated, this view rejected mathematical objects, the proof of whose existence depended on an infinite number of steps. Any existence proof that implicitly depended on a greater than finite number of steps could not be regarded as constructive since the existent, even in principle, was

10. Russell's Paradox was put forward in 1902. Some sets are not members of themselves, for instance, the set of all women. Other sets are members of themselves, such as the set of all things that are not women. If one considers the set consisting of all sets who are not members of themselves, one finds the paradox that if it is a member of itself, then it is not, and vice versa. Antinomies such as these gave rise to restrictions on the use of general properties to define sets.

unattainable.[11] Logically, this meant the rejection of much that classical mathematics took for granted, including the Principle of the Excluded Middle, Cantor's theory of infinite sets, and many existence proofs. To Hilbert, who considered such concepts central, the denial of all this to mathematics was akin to "prohibiting the boxer the use of his fists" (Reid, 149). By 1922, his agitation was heightened further by the degree to which intuitionism seemed to be taking hold among younger mathematicians, such as Weyl, and, turning from physics, he threw himself at his work on axiomatics with renewed vigor. Mathematics, if it were to remain intact, had to establish deductions with the same certitude that existed for the arithmetic of whole numbers "where contradictions and paradoxes arise only through our own carelessness" (quoted in Reid, 176).

Meanwhile, Hilbert's colleagues were contending with the proliferation in physics, which by 1926 had become even more notable. Heisenberg's new theory of quantum mechanics was shown to be explicable in terms of matrix methods by Max Born. Then Schrödinger, at Zurich, constructed his wave mechanics which, while it led to the same results as Heisenberg, proceeded from an entirely different base. The two were soon mathematically reconciled by Courant, using, to a great extent, Hilbert's earlier work on integral equations and infinitely many variables (Hilbert Space). These radical developments in physics challenged prevailing views in mathematics. The theory of relativity deeply questioned many concepts that were central to classical mechanics, such as absolute space and time, simultaneity, and so on. Quantum theory, more importantly, threw mechanistic determinism into disarray by demonstrating the impossibility of knowing simultaneously both the position and velocity of a particle, without which its future evolution could not be predicted. Such basic contradictions called for fundamental changes in the mathematics used, to the chagrin of classical mathematicians such as Poincaré and Volterra, who believed in the underlying continuity of physical events and the possibility of their representation by the infini-

11. Reid (149) offers the following example: consider the statement A, "There is a member of the set S having the property P." If S is infinite, each member of the set cannot, even in principle, be examined to verify the statement. Brouwer rejected such existential statements for infinite sets. Now the Principle of the Excluded Middle says that if A holds then ~A does not hold, and vice versa: there is nothing in between. Brouwer argued as follows: if in S above one finds one element showing property P, then the first alternative is substantiated. If, however, one does not, then the second alternative is still not substantiated, nor has the middle been excluded.

tesimal calculus and the theory of differential equations. As Ingrao and Israel (1990) point out, the emphasis shifted from *mechanical* analogy to *mathematical* analogy. Mathematics was to be used for its form, to provide a language unifying theories, rather than to mechanically describe physical processes.

If strong tremors, therefore, were being felt throughout the mathematical world at this time, the Göttingen to which von Neumann was first directly introduced in 1923 was at the epicenter. His contributions for the next few years reflected the concerns of the time and the place. Hilbert's twin concerns about the foundations of mathematics and the axiomatization of mathematical physics greatly influenced him (see Heims 1980, Goldstine 1972). Working with Lothar Nordheim, Hilbert's assistant in physics, von Neumann undertook the axiomatization of Heisenberg's work and then proceeded to develop further the notion of the Hilbert Space to provide a fuller mathematical basis for quantum mechanics. His seminal work on the axiomatization of physics appeared in the form of three articles in 1927, three in 1929, and their condensation into a book, *Mathematical Foundations of Quantum Mechanics* (1932). In 1927, influenced by Hilbert's concern with the foundations of mathematics, he also published a paper conjecturing that all analysis could be proved consistent. At some point in 1926, von Neumann produced his proof of the minimax theorem which, not surprisingly, was overshadowed by his contemporaneous work (see Heims 1980, 56). The source of his interest in games thus remains something of a mystery. However, his use of the axiomatic approach is entirely in keeping with the Hilbertian ethic with which he was fully imbued, and the notion of chance, made central through probabilistic play, is consonant with the indeterminism at the basis of quantum mechanics: "Chance . . . is such an intrinsic part of the game itself (*if not of the world*) that there is no need to introduce it artificially by way of the rules of the game: . . . it still will assert itself" (von Neumann 1928b, 26; emphasis added). As noted above, there are two publications by von Neumann dealing with the minimax proof (1928a, 1928b), the first of these being the communication to Borel presented by the latter to the Académie. In it, von Neumann refers to Borel's work in this vein since 1924 and claims he has solved the problem of the existence of a best way to play the two-person zero-sum game. He points out that he reached these results independently: "M'étant occupé indépendamment avec le même problème" (1928a, 1689), and that the full proof is forthcoming soon as "Zur Theorie der Gesellschaftsspiele" in

Mathematische Annalen, the Göttingen journal (which had accepted it for publication by this time).

The second, von Neumann 1928b, appeared later that year, primarily containing a long and difficult proof of the existence of an equilibrium value for the two-person, discrete game, based on functional calculus and topology. The paper reveals little about whose work, if anybody's, von Neumann draws upon and there is no reference to the past except in two footnotes. The first of these says that the paper had been presented in shorter form in December 1926 to the Göttingen Mathematical Society.[12] The second states that while finalizing the current paper he "learned of the note of E. Borel in the Comptes Rendus of Jan. 10, 1927" (25n). In the paper, the concept of a game is completely axiomatized and two examples are offered of zero-sum games with solutions only in mixed strategies.[13] Among the situations that can be regarded as games of strategy are roulette, chess, baccarat, bridge, and "the principal problem of classical economics: how is the absolutely selfish 'homo economicus' going to act under given external circumstances?" (13).[14] Von Neumann also treats the three-person zero-sum game, showing how the possibility of coalition formation introduces a measure of indeterminacy, or "struggle," into such games. In preliminary remarks on games with more than three players, he broaches the topic to which von Neumann and Morgenstern (1944) would later devote much space. Without calling it such, he introduces the characteristic function, a "system of constants" describing "the sum per play which [each] coalition of the players . . . is able to obtain from the coalition of the other players," and conjectures that "*the complex of valuations and coalitions in a game*

12. This was undoubtedly the Göttingen Mathematical Club, "the highpoint of the mathematical week . . . during the 1920's" (Reid 1970, 168). This was an informal gathering where faculty and students gave talks on their recent work.

13. The two games illustrated are matching pennies (here the payoffs are given but the name is not used) and morra (also called "paper, stone, scissors"). The latter is one of the games considered by Borel in his 1924 note.

14. This is the first indication of von Neumann's interest in economic matters. Nicholas Kaldor (in Dore 1989) recalls von Neumann expressing such an interest in 1927, not long after his Berlin appointment. They met while returning on holiday to Budapest as young scholars working abroad. Kaldor suggested he read Wicksell's *Value, Capital, and Rent*, which provided an introduction to Walras and utilized Böhm-Bawerk's capital theory. Von Neumann, on reading this, criticized the Walrasian system, saying that it permitted negative prices. Kaldor speculates that this may have prompted von Neumann to write his 1937 paper on economic growth, and that furthermore, von Neumann's interest in economics may well have grown out of his interest in game theory.

of strategy is determined by these . . . constants alone" (40–41; emphasis added). If this is the case, then "we have brought all games of strategy into a natural and final normal form" (41). In conclusion, von Neumann adds that a later publication will contain numerical calculations of a simplified poker and baccarat, the results of which are corroborated by the well-known necessity to "bluff" in poker.

The French Connection

I observed above that von Neumann's communication with Borel was minimal, peremptory, and more in the nature of a rebuff than an invitation to exchange ideas. Under these circumstances, it is interesting to inquire as to what subsequently happened in the 1930s to the latter's work on games. Did he, for example, incorporate von Neumann's result into his analysis? Or did he take up the framework suggested in the 1928 paper? However, even these questions suggest too smooth and seamless a conception of the history of ideas: what happened, briefly, was that Borel seems to have gradually lost interest; his student, Jean Ville, took up the cudgel, doing some further original work; and one of his contemporaries, René de Possel, brought the good news in popular form to the French intelligentsia. *And all this went by, it appears, unbeknownst to von Neumann.*

In 1936, mathematician René de Possel wrote what might be regarded as the first popularization of von Neumann's minimax theorem. This came as part of a series of monographs on original contributions to science, philosophy, literature, and art, produced by the Centre Universitaire Méditerranéen de Nice and edited by the poet Paul Valéry. These were clearly intended for the well educated but nonspecialist reading public. De Possel's booklet is a forty-page description of the analysis of popular games, in which von Neumann's theorem is presented as the culminating achievement to date. Following Borel 1924, games are divided into those based on pure strategy, those based solely on chance, and those involving "ruse," or bluff, where a player can gain an advantage by knowing the opponent's intentions. De Possel discusses various examples. For example, in the game of "batonnets" each player draws a number of matches from a pile until it is depleted. No more than a certain number may be drawn each time and the last person to draw wins the game. Based on the number of matches in the pile, an optimal strategy exists for at least one of the players, demonstrable by backward induction. Various versions of this purely strategic game are discussed. Roulette is presented as a game of pure chance. In this context, he explains the idea of the martingale, or how best to spread one's bets across several rounds, given the capital at one's disposal and the unfolding pattern of wins and losses. The ubiquitous "Baccarat du Bagne," or "scissors, paper, stone," is presented as an example of a "social game," which combines strategy, chance and "ruse," and von Neumann 1928b is invoked to show how minimax play is optimal in that it eliminates the risk of the opponent guessing one's intentions. While the theorem itself is "too technical to be reproduced here" (39; author's translation), it is clearly the booklet's central feature, and von Neumann is honored as "the first to seek to penetrate the mechanism of play from such a general perspective" (5; author's translation).

However, no such credit is afforded von Neumann in Borel's work of the same period. In 1938, as part of his extended treatise on probability, Borel contributed a volume on "Applications to Games of Chance." This is a version of his course on the topic at the École Normale in 1936–37, written and edited by his student Jean Ville.[15] Following his by now standard approach, Borel initially examines at length dice and simple card games, where chance is the dominant feature and the analysis is confined to combinatorial probabilities. Then, in a chapter called "Games Where Psychology Plays a Fundamental Role," he synthesizes and extends his work of the 1920s. Again, both finite and infinite games are analyzed. In the case of the former, the usual suspects are featured, for example, "Heads-or-Tails" and "Baccarat du Bagne," and the determination of optimal mixed strategies discussed in detail. Remarkably, however, *no* mention is made of what had become, for de Possel at least, the "théorème fondamental," that of von Neumann (1928b). This can only be regarded as an act of deliberate omission by Borel. Indeed, he devotes most attention to continuous games. For example, in the game where two players must each choose a point on a circle, some fair criterion determining the winner,[16] simple integration is used to construct the distribution that each should optimally apply in selecting a point; the density is uniform so all points are afforded the same probability. This is extended to the choice of three points from a continuum by each player,

15. It was, and to some extent still is, quite common for some French professors to have a *protégé* take careful class notes for later publication.

16. On a fixed circle, a chosen point wins if the other choice lies within an arc of length p, in the counter-clockwise direction. Otherwise, by definition, the other choice wins.

and the same mathematical justification is provided for the intuition of random play. Once again, Borel considers situations to which these principles are connected, in both the military and economic arenas. For example, he suggests that the analysis might be applied to the allocation, by opposing armies, of their respective forces to a limited number of common strategic points, or to the problem of how two merchants, wishing to sell equal stocks of similar goods, should distribute their available discounts across the goods, given that they are competing for customers.[17] In a final chapter, a simplified poker is analyzed, showing the importance of bluffing. It is possible that this was prompted by von Neumann's earlier indication to Borel that he had done such work.

Following the above, and occupying but 9 of the book's 120 pages, is a note by Borel's student Jean Ville, which, in the purely mathematical sense, was to have a greater subsequent impact on the consolidation of the theory than the rest of the book. This, of course, is Ville's construction of the first elementary proof of the von Neumann theorem for finite games and his own extension of this to show that a simple infinite game also has a value. Ville's proof is partly topological and rests on a theorem on linear forms in nonnegative variables. Compared to von Neumann's earlier contortions, it is positively elegant.

Consider p linear forms in n variables:

$$f_j(x) = \sum a_{ij} x_i \quad (j = 1 \ldots p; i = 1 \ldots n)$$

with the property that, whatever the nonnegative values of x_i, there exists among the f_j at least one that is nonnegative. Then, there exists at least one system of nonnegative coefficients $Y_1 \ldots Y_p$, $(\sum Y = 1)$, such that $\sum Y_j f_j$ is nonnegative for all nonnegative values of variables x_i. Having proved this, Ville then establishes a corollary: in the same system, if, whatever the nonnegative values of x_i, there exists at least one f_j no less than φ, itself a linear form in the same variables, then there exists a linear combination,

$$\psi = Y_1 f_1 + Y_2 f_2 + \cdots + Y_p f_p \quad \text{with } Y_j \geq 0, \ \sum Y_j = 1,$$

such that

$$\psi \geq \varphi \quad \text{at every point } x \text{ (where } x_i \geq 0, i - 1 \ldots n).$$

17. It is this military example that would later feature as the Colonel Blotto game, first appearing in Morse and Kimball 1946 and then developed at RAND in the late 1940s, ostensibly to consider the global allocation by the United States and Soviet Union of their respective armaments. See Dresher 1961.

Having defined a game and shown each player's expectation, conditional on the opponent's strategy, to be a system of linear forms, Ville simply invokes the above theorem and corollary to show the existence of a value that represents at once an assured upper limit on the minimizing player's expectation and a lower bound on that of the maximizing player. He then, for the first time, draws the infinite game into the compass of the minimax concept. Considering a simple game where each player chooses a point from the unit interval, player A's expectation is given by

$$\int_0^1 \int_0^1 K(x, y) dF d\Phi,$$

$K(x, y)$ being A's payoff given choice of points x and y; F and Φ being the probability distribution functions applied to the strategies chosen by A and B, respectively. Ville shows that if $K(x, y)$ is continuous in x and y, in the closed domain $0 \leq x \leq 1, 0 \leq y \leq 1$, then this game too has a value. This he does by establishing this infinite game as a limiting case of a finite game with very many strategies.

Following this, Borel offers some observations on his student's contribution which unambiguously reveal his doubts about the entire theory. Having thanked Ville for illustrating von Neumann's "important theorem," he says, "It appears essential for me to indicate, however, to prevent all misunderstanding, that the practical applications of this theorem to the actual playing of games of chance is, for a long time, unlikely to become a reality" (Borel 1938, 115). Actual games are exceedingly complicated, he says, containing many coefficients and equations. Even if one could simplify a game sufficiently to the point where such calculations were possible, the advantage of playing according to the above theorem are only attainable on average, after a great number of rounds. And even taking account of the experienced recommendations of players to locate what might be regarded as reasonable strategies, "there still remains such a great number of variables that even the task of writing the equations, not to mention that of solving them, appears absolutely insurmountable" (115).

In Borel, one senses a respect for games, many of which are old and have been played for generations. In many such games, a consensus may emerge on what constitutes good play and thus what the novice should be taught. However, no sooner has such agreement been reached than the better players take advantage of it to introduce newer, more successful ways of playing. This is what makes games interesting. Indeed, it often goes full circle, with today's innovators reviving ways of playing that were once considered revolutionary but then abandoned. Can it ever be hoped, he asks, that this natural evolution will approximate the impracticable solution of a system of equations that completely describes the game? "This, I must admit, seems highly doubtful to me, and, anyway, if it happened for a particular game, it is almost certain that the game would soon be abandoned for a more complicated one" (116). Furthermore, even where ideal play involves a probabilistic element, it is very difficult not to follow some regularity when actually playing. This is particularly important in bridge, for example, where probabilistic play intended to defeat one's opponents may well mislead one's partner also! "All these remarks . . . are obvious to anyone with some experience in games. Perhaps they will make clear, to those uninterested in games, how enjoyable games are as leisurely distraction, at the same time showing to those who would wish to turn games into an occupation, how futile is the search for a perfect formula which is forever likely to elude us" (117). These remarks effectively signal the end of Émile Borel's active contribution to game theory.[18] They both constitute the first criticisms of the minimax idea, and suggest his unwillingness to sacrifice the mystery and delight of games for an elegant but inapplicable mathematics. His is a refusal to take it all too seriously.

Von Neumann in the 1930s

After a one-semester visit to Princeton in 1929 to lecture on quantum theory at the invitation of geometer and topologist Oswald Veblen, von Neumann alternated between there and Berlin, continuing his work toward the *Mathematical Foundations of Quantum Mechanics* (1932). During this period, his interest in economics was further stimulated and, in fact, the minimax idea resurfaced in the context of his 1937 growth model. At Menger's seminar at the University of Vienna, a sequence of papers presented between 1932 and 1937 dealt with equilibrium and growth in Cassel-type models (see Weintraub 1985, 72–78). These were usually published the year following in the *Ergebnisse* (translated, *Proceedings*), edited by Menger. Karl Schlesinger (1932) offered a reformulation of Cassel using inequalities and set forth a model, without mathematical analysis. Following this, Wald, Menger's student, proved the existence of a solution to Schlesinger's model (the first existence

18. Subsequent publications merely reiterated his earlier work, for example, Borel 1950, Appendix 1.

proof) and modified this further in 1936. Von Neumann (1937) removed the distinction between primary factors and outputs: all goods are produced. Rather than emphasize production of single goods, he uses processes; one process may produce multiple outputs and each output may result from different processes. He then characterizes the equilibrium rate of growth as the saddle point of a function relating the input and output matrices, its existence proved using Brouwer's fixed point theorem.[19] Apparently, von Neumann had presented the paper at the Princeton Mathematical Society in 1932 and, unlike the other papers in the series, did not present it to the Vienna seminar.

The 1930s, therefore, saw the affirmation of von Neumann's interest in theoretical economics. Among economists, it is for his contribution on the growth model during this period that he is remembered. It is also true, however, that the "pure" mathematics of games was on von Neumann's mind during this time. In April 1937, in its mathematics section, *Science News Letter* reported a talk by von Neumann at Princeton on what for him was "a mere recreation," his analysis of games and gambling. Apparently, he spoke about "stone-paper-scissors," showing that by "making each play the same number of times, but at random, . . . your opponent will lose in the long run." Also parsimoniously reported are his observations on the probabilities of making particular plays in both dice and a simplified poker. Two-and-a-half years later, in November 1939, von Neumann was planning a visit to the University of Washington, where he was to spend part of the upcoming summer semester as Walker-Ames Professor in mathematics. In a letter to the department, he suggests possible topics for his lectures, including quantum theory, operator theory, groups, and the "Theory of Games." On the last he says, "I wrote a paper on this subject in the Mathematische Annalen 1928, and I have a lot of unpublished material on poker in particular. These lectures would give a general idea of the problem of defining a rational way of playing" (von Neumann Papers [VNP], Container 4, File 3, Personal Correspondence 1939–40).

Finalizing matters four months later, in March 1940, he indicated that he would give three evening lectures on games:

19. Despite apparently taking Cassel's work as a starting point, the paper contains no reference to the related work of Schlesinger or Wald. Wald and Menger, however, accepted the paper for publication without exercising their editorial prerogative. Arrow (1989) notes that "von Neumann's lack of references is in general a source of difficulty in reconstructing the evolution of ideas" (17).

1. The general problem. The case of chess.
2. The notion of the "best strategy."
3. Problems in games of three or more players.
General remarks (File 3).

The extent to which Seattle's mathematicians were stirred by von Neumann's still-quirky ideas is unknown. What is clear, however, is that just prior to this sojourn he had captured the imagination of one distinctly nonmathematical economist who had moved to Princeton from Vienna two years previously, Oskar Morgenstern.

Morgenstern's Early Career

Born in Silesia, Germany in 1902, Oskar Morgenstern moved at twelve years of age with his family to Vienna where, in 1925, he obtained his doctorate with a thesis focusing on marginal productivity.[20] Following three years visiting London, Columbia and Harvard Universities, Paris, and Rome as a Rockefeller Fellow, he was appointed Privatdozent at the University of Vienna in 1929. Morgenstern's habilitation thesis and first book, *Wirtschaftsprognose* (translated *Economic Prediction*, 1928), focused, in the Austrian tradition, on epistemological difficulties in economic forecasting: the difficulties of "knowing" when "other wills," other "economic acts" may "interfere with, or enhance, one's own plans" (Morgenstern 1976, 806). During the 1930s, his work concerned such issues as the business cycle, methodology, and the treatment of time in economic theory. He also had some work published in the United States and both George Stigler and Frank Knight were familiar with his work. His main professional activity was the directorship of the Vienna Institute of Business Cycle Research and he also acted as editor of the *Zeitschrift für Nationalökonomie*.

It is easy to look back at his work and pick out those elements that are most congruent with what later appeared in the *Theory of Games*. However, I believe his position as coauthor with von Neumann is better understood by focusing not on particular incipiently "game-theoretic" concepts he may have alluded to in his earlier career, but rather on his

20. For the fullest extant treatment of the life and work of Oskar Morgenstern, the reader is referred to Rellstab (1991, 1992), both of which are part of a larger project. In gaining an understanding of Morgenstern, I have benefited tremendously from conversations with, and the generosity of, Urs Rellstab.

general position as an arbiter of ideas, an intermediary between theorists in disparate fields, and one ultimately most capable of giving his energy to penetrating criticism rather than alternative theoretical construction. First of all, Morgenstern was primarily an outsider: neither intellectually nor psychologically did he fit comfortably into any particular group or school of thought. He attended the meetings of the Vienna Circle and was much taken by the philosophical flux centered on Schlick and Carnap, but he was not a philosopher, and he discussed these developments at a distance. Neither was he a mathematician, and while he attended the Menger colloquia and debated issues in mathematics and their relationship to the social sciences, his relationship to mathematics remained akin to that of the impresario to the music.[21] Second, Morgenstern's intellectual interests, while varying in depth, were certainly broad, and he consistently attempted to relate disparate developments in philosophy and mathematics to economic theory. In "The Time Moment and Value Theory" (Morgenstern 1935b), the work of Karl Menger and Moritz Schlick in logic forms the background for his critique of the treatment of time in neoclassical economic theory. Similarly, in "Perfect Foresight and Economic Equilibrium" (1935a) and "Logistics and the Social Sciences" (1936), he contends that only by incorporating recent developments in logic can economics achieve what he saw as the necessary level of rigor. The need for mathematical rigor he emphasized incessantly in both his public remarks and personal reflections: "When I ask myself what I consider to be my main duty working on economic problems, it is the introduction of truly exact reasoning and truly exact methods" (OMPD, Diary, Box 13, 19 April 1936). Broadly familiar with the developments in mathematics of Russell, Hilbert, and their contemporaries, he felt that the implications for economic theory were enormous. Indeed, he sought to overcome his deficient training in mathematics by taking private tutorials with Abraham Wald, whom he appointed as researcher at his institute, but he remained personally incapable of taking the theoretical steps that he himself envisioned.[22] Indeed, if any irony characterizes Morgenstern's career, it is that, in his

21. I believe his authorship of the *Theory of Games* created a false impression about his capabilities in this regard. In presentations on the subject for years afterwards, Morgenstern felt obliged to preface his remarks with the observation that they would be nontechnical. Only by 1964, as the dusk of his career approached, did he say bluntly, "Mathematical intuition is a very important thing. I wish I had some" (quoted in Mensch 1966, 100).

22. Morgenstern hired Wald at the suggestion of Karl Menger, Wald's teacher.

continuous agitation for mathematical rigor in economics, he was ultimately calling for a theoretical approach in which thinkers of his own kind would have increasingly little place.

In 1938 he visited the United States with the support of the Carnegie Endowment for International Peace and soon found himself unable to return, having been dismissed by the Nazis in his absence. When offered a three-year appointment in political economy at Princeton, he accepted. His decision to stay in the United States was not one reached suddenly, however, for two years earlier he had written to Frank Knight,

> The idea of being a professor in a large and reputable American University appeals to me very much. You know very well, that I have a high regard for research in social sciences in the States and I have tried to transplant a good deal of the American method to Vienna. . . . Anyway, I wanted to make clear to you that my reaction to your question whether I would like to come to the States is possitive and I wish you would let me know if there are any further developments. (OMPD, Box 6, Correspondence 1928–39, Knight, F., 6 April 1936)

One should note, however, that even though he had been made aware of von Neumann's 1928 paper on games by the Czech mathematician Eduard Cech, he did not read the work until he met von Neumann in Princeton at the end of the 1930s.

Collaboration at Princeton

By 1938, therefore, both Morgenstern and von Neumann were at Princeton. One an economist, the other primarily a mathematician, they differed in many ways, but their common situation likely eclipsed differences that back in Vienna or Göttingen might have been more significant. The two always spoke German together, even when writing in English, and there were dinners and conversation with Einstein, Bohr, and Weyl (Morgenstern 1976). A glowing account of their collaboration on the *Theory of Games and Economic Behavior* (1944) is given in Morgenstern (1976). This reminiscence, while it may capture Morgenstern's nostalgia as he looked back at the high point of his career, is at odds with the version of events actually recorded in his personal journal at the time of his work with von Neumann during World War II.

My concern here is to understand how their book reflects the different interests of the two authors and how it relates to the explorations discussed above. As mentioned previously, von Neumann and Morgenstern met in the fall of 1938, shortly after Morgenstern arrived at Princeton. Despite several further meetings in the interim, not until April 1940 did they actually discuss games and economics, by which time, we have noted, von Neumann had already arranged his lectures on games for Seattle. During these discussions, von Neumann, we are told, read and praised Morgenstern's "Time Moment in Value Theory." While away, von Neumann did further work on games, moving on to consider four-person games. On his return, in the fall of 1940, von Neumann began to write a paper in two parts synthesizing his work on game theory to date (1940, 1941). This constitutes the theoretical framework of what became the *Theory of Games*. From the beginning of the first part, "General Foundations," the emphasis is clearly on games with more than two players: two-person theory is given short shrift, being quickly used to motivate the notion of the (characteristic) set function v(S). The concepts of strategic equivalence and (in)essentiality are defined, and the three-person game is illustrated in detail, showing how its solution is a system of possible "apportionments" (imputations). The stability evident in this case is considered in the general n-person game and a complete definition is given: no two valuations in the solution dominate each other and every external valuation is dominated by at least one solution member. This, of course, became known as the "Von Neumann–Morgenstern Stable Set" after 1944. In addition, the concept of (in)essentiality is discussed and a graphic illustration of the three-person game provided. All the aforementioned is seen as a prelude to "our ultimate goal, i.e., to find the solutions of all games with $n \geq 4$" (1940, 25). The second part extends the coverage to include non-zero-sum situations, showing how the analysis remains essentially the same, and, adopting a more formal set-theoretic notation, proves some simple theorems on stability and discusses decomposition of games (the properties of games when considered together versus separately). Throughout, the presentation is dense and rigorous and without discussion of economic or any other applications.

While von Neumann was writing this extended paper, however, he was listening to criticisms of conventional economic theory by Morgenstern, who was then preparing his scathing (1941b) critique of Hicks's (1939) *Value and Capital*. Provoked by this, in May 1941 he asked Morgenstern to write a paper illustrating his basic thoughts on economic theory. Within a month, this yielded "Quantitative Implications of Maxims of Behavior" (Morgenstern 1941).

If von Neumann's work above is impressive for its relentless mathematical rigor, the paper by Morgenstern is equally impressive for different reasons. Devoid of mathematics, it offers a skeletal social theory that is both methodologically individualistic and cognizant of the importance of social interaction. His concern, he states at the outset, is to construct a "theory of society," but we will limit ourselves to economics since "there alone the beginnings of a theory of social behavior are discernible" (1). To this extent, a "maxim" may be regarded as a complex plan. Morgenstern distinguishes between two types of maxim. Unrestricted are those whose success or failure is independent of their adoption by other individuals, while restricted are those who are so dependent. A successful economic theory, that is, one that enables prediction, will have to recognize that individuals make decisions whose outcome depends on whether or not similar decisions have been made by others. This is the core of Morgenstern's message.

In the process, Morgenstern conveys to the reader an idea of how he views knowledge, the individual, society, and economic theory. There exists an objective mechanism which underlies the evolution of economic phenomena. Individuals, having less than perfect foresight, have incomplete knowledge of "the facts," and thus act with subjective, rather than objective, rationality. In addition, because all decisions take place in time, and because time changes maxims from unrestricted into restricted ones, its introduction into any theory is "essential" (2). This is emphasized repeatedly by Morgenstern, for example, "Restricted maxims will be applied in succession and have therefore a dynamic character" (21a). The quantitative effects of other individuals' behavior on one's application of restricted maxims may be positive or negative. For example, one's decision to withdraw deposits in order to protect them from bank failure will depend on the number of others making a similar decision. In cases such as this, the rift between subjective and objective rationality may be very large indeed. This raises the importance of the "institutional setting" in which economic decisions are made: regulation may help overcome information deficiencies, thereby reducing these interaction effects. Interestingly, Morgenstern claims that the case for government intervention is made stronger when it serves to enhance inadequate private information, making easier the pursuit of restricted maxims. This, he says, is something not recognized by the laissez-faire doctrine. In general, however, "social progress [might] be presented as a gradual shift from restricted to unrestricted maxims" (6a).

The document is manifestly positivistic: the entire theory is predicated on the gap between the true mechanism of the economic world and our knowledge of it, both of which are completely quantifiable. The only suggestive mathematics in existence is "the theory of games by J. von Neumann," referring to the paper discussed above, but, perhaps in an attempt to provoke the latter to further efforts, he adds that even this does "not take the problems into consideration which have been described above" (22).

By September 1941, the two decided that their papers (above) should be combined to form a small volume. In an October letter to Frank Aydelotte, director of the IAS, von Neumann wrote, "We hesitated for some time whether to publish it as a paper (in one of the economics periodicals) or as a book. We are now inclined to do the latter since this would free us of limitations in space which would be rather troublesome."[23] It ended up, three years later, as the 635-page tome with which we are now familiar, "a big book, because they wrote it twice, once in symbols for mathematicians and once in prose for economists" (David Blackwell quoted in Albers and Alexanderson 1985, 27). In the introductory chapter, "Formulation of the Economic Problem," Morgenstern offers both a critique of accepted neoclassical theory and a reworked version of the alternative conceptual approach broached in his "maxims" paper. While the wings of Morgenstern's imagination have been clipped this time by the need to have his theoretical hopes conform to the "actually existing mathematics," the grandiloquence remains. The importance of interdependence is argued: when we move from the Robinson Crusoe world to a social group, qualitatively new features enter the picture; these features require a mathematics better adapted than the traditional differential calculus. Throughout, physics is presented as a benchmark in scientific progress which economics should strive to attain. Gone is the emphasis on time and dynamics: "Our theory is thoroughly static" but, as the experience in physics has shown, "it is futile to try to build [a dynamic theory] as long as the static side is not thoroughly understood"

23. Dated 6 October 1941 in VNP, Container 32, File 90, Theory of Games and Economic Behavior. Aydelotte responded, "I was keenly interested in your suggestion that the mathematical theory of games might demand the development of new branches of mathematics running parallel to the impulse which physics gave to mathematics in the seventeenth century. If you can get the mathematicians to take as much interest in that side of your paper as the economists are sure to take in the economic aspects, you would indeed be starting something of first-rate importance" (9 October 1941). As we shall see below, for the next twelve years or so, game theory was a distinctly mathematical, rather than economic, affair.

(44). The rest of the text is a lengthy elaboration of the theory, centering largely on games of three or more persons, with Morgenstern's economic examples appearing in chapter 11.

How do von Neumann and Morgenstern situate their book with regard to the earlier work considered in this article? While Steinhaus, as we know, remained incognito, and therefore could not have been taken into account, even in principle, it is remarkable that none of Borel's work before 1938 is mentioned. The latter volume, furthermore, is briefly referred to twice. The first concerns poker, where Borel's related work is characterized in a footnote as "very instructive, but . . . without a systematic use of any underlying general theory of games" (186, n. 2). The second mentions the first elementary proof of the minimax theorem by Ville in the same volume. The proof used by von Neumann, based on the theory of convex sets, is offered as a further, simplifying step in this process of elementarization: in general, the text presents itself as a revolutionary one, whose relationship to earlier mathematical and economic ideas is one of contrast rather than continuity.

The Period to 1944: Overview

The period leading up to the publication of the *Theory of Games* displays historical features against which the subsequent development of game theory may be juxtaposed and without which it cannot be understood. If one term can possibly capture the twenty-five or so years considered above, it is "fragmentation." From 1921 until the collaboration of von Neumann and Morgenstern, the three interested parties worked in relative isolation. In Poland, Hugo Steinhaus seems to have remained disconnected from and unaware of the contemporaneous writings of both Borel and von Neumann.[24] The relationship between the latter two bears the hallmarks of a certain "standoffishness," if not outright academic antagonism. At no stage, despite their mutuality of interest, did either show

24. Although its broader significance for our story remains a matter of speculation, both Borel and Steinhaus did have one thing in common: a strong interest in naval matters. We have already noted Borel's involvement as Minister of the Navy. Commenting on his own 1925 paper, Steinhaus (1960) wrote, "I was especially interested in naval pursuit. After having found the concepts of minimax and maximin I was well aware that the minimax time of the pursuer is longer or equal to the maximin time of the pursued, but I did not know whether they are equal in all similar games. Consequently, I called 'closed' the games for which there is equality and 'open' the other ones. My pupils in Poland have adopted later this terminology (thus the pursuit of one ship by two is *closed*, and it is *open* when there are three pursuers)" (108).

a willingness to cooperate or exchange ideas. Nor is this particularly surprising. There is a natural spirit of competition felt by mathematicians with regard to their work. Priority debates, for example, reflect this sense of pride and the desire to be given credit for initiative and originality. In von Neumann's case, this natural tendency may have been magnified by two other features. First, as noted above, he was considerably younger than Borel and, while his genius was being recognized, he no doubt relished the prospect of outsmarting somebody of Borel's stature. Second, it reflected the longstanding rivalry in mathematical matters between Göttingen and Paris which, in turn, was in keeping with the volatile nature of Franco-German diplomatic relations.[25] In the work of both Borel and von Neumann, from 1928 onwards, references to the contribution of the other were scant or nonexistent. Despite the fact that each clearly drew on the other's work to some extent, whether in the choice of examples or the basic framework, relations between them remained distant and disconnected. At no stage could anybody have spoken of a game theory "community."

Even in their collaboration on the book during the early 1940s, von Neumann and Morgenstern worked in isolation. For example, neither seems to have been aware of the work in France: in December 1941, Morgenstern *accidentally* discovered Borel's volume (1938) containing the elementary minimax proof by Ville. Neither was von Neumann aware of this work (see Morgenstern 1976, 811; Rellstab 1992). This of course is not too surprising, given the political situation. Even at Princeton, however, they were quite isolated. Through 1941–42, von Neumann gave a few related lectures which were not well attended, and the economics department at Princeton, to Morgenstern's dismay, remained aloof. He speculates that this kind of work may have somehow been inappropriate to the conditions of war then obtaining.[26] That game theory was shaped at Princeton in virtual isolation from the mathematics and economics communities there is also borne out in other accounts. Ted Harris, later to be head of mathematics at RAND, received a Ph.D.

25. Reid's 1970 anecdotes endorse this. For example, the 1909 visit by Poincaré to Göttingen "was an unwelcome reminder that the mathematical world was not a sphere, with its center at Göttingen, but an ellipsoid" (120). Further, in 1917, when Hilbert wrote a memorial for the deceased Gaston Darboux, his house was besieged by a mob of angry students demanding that "the memorial to the 'enemy mathematician' be immediately repudiated by its author and all copies destroyed" (145).

26. Ironically, as we shall see, it was to be the consideration of war that would motivate the theory's early development.

in mathematics in 1947 and recalled that his first encounter with game theory was his chance discovery of the *Theory of Games* in Princeton's bookshop, just before leaving the university. Harris had been unaware of this development during his career at Princeton and had seen von Neumann speak only on ring operators and other topics (Harris 1990). Albert Tucker, student of Solomon Lefschetz and then professor of mathematics at Princeton, similarly did not become interested in game theory until 1948 when George Dantzig invited his participation in a trial project on linear programming being supported by the Office of Naval Research (ONR; Albers 1985, 343). Likewise, Sam Karlin, a fellow graduate with Harris, and soon to be a key game theorist at RAND, consulting from the California Institute of Technology (Cal Tech), had had no exposure to game theory at Princeton. The *Theory of Games*, he argues, appeared to have had little effect on a Princeton mathematics department oriented primarily toward topology and analysis; indeed, Karlin suggests that game theory came into existence there only after he *left* in 1947 (Karlin 1990).

Thus, the *Theory of Games and Economic Behavior*, when it emerged, was without a "natural" audience. Though directed in its rhetoric toward economic theorists, its central ideas were equally novel in the mathematical sense. While the authors hoped that the impact would be greatest in economic theory, they also surmised correctly that some time would pass before the "game" idea became common currency. The second part of this article illustrates the emergence of a game theory community. For the first time, there appeared an extended group, among whose members the exploration of the mathematics of strategic interaction became a *modus vivendi*. What is of particular interest to economists today is the curious role played in the whole affair by the *Theory of Games*. The book certainly signaled to the broader audience in mathematics and economics that a theoretical innovation had taken place. However, a community of "game theorists" emerged *not* through their exploration and elaboration of the "great book," but rather as a result of the role of the older minimax idea in World War II. Postwar mathematicians came to grips with game theory by extending the earlier work of von Neumann and Ville on two-person games, *not* the characteristic function touted as the theoretical centerpiece of the *Theory of Games*.

Part 2: Postwar Consolidation

The significance of wars in determining the direction of research and, therefore, the form of "discovered knowledge" is enormous. In this regard, the effect of World War II was unprecedented. The application of scientific methods to conflict not only destroyed lives and cities more "efficiently" than ever before, but had fundamental repercussions on the way science would be pursued when the conflict ended. And this applied to mathematics and economics as much as to the natural sciences. Shortly after the end of the war, interest in game theory had spread to other mathematicians besides von Neumann. It started to gain respectability as an area of applied mathematics, support for further research in the area was forthcoming, and a self-reproducing community appeared. The entire process represented the stabilization of the mathematics of games. This occurred through the demonstration of the links between games and what was already known, for example, in other areas in mathematics or what had recently been learned in the area of military strategy. It is somewhat ironic that while the grander ambition of the *Theory of Games*, particularly as expressed in Morgenstern's introductory chapter, was temporarily shelved in the process, von Neumann himself was a fundamental force in the entire movement.

I first look at the role of academics in wartime research and then concentrate on two loci of particular interest: the Operations Evaluation Group, attached to the Navy, and the Statistical Research Group, attached primarily to the Air Force and part of the Applied Mathematics Panel. It was in the former that game theory found its first military application, while the latter would form the core of a group of postwar mathematicians centered at Project RAND, devoted to research on games. The work at RAND reflected several influences: the continuation of research on military strategy as demanded at the end of the war; the exploration of games from the perspective of von Neumann's new interest, computation; the exploration of the theoretical links between games and linear programming on the one hand and statistical decision theory on the other. Out of this milieu came the first textbook, the first layman's version of game theory, and the personnel who would carry the theory back to campus mathematics departments, as they were revived by the ONR.

Academics and War

The German invasion of Poland in 1939 highlighted the fact that there were "grave shortcomings in the organisation of science for war" in the United States (Baxter 1946, 11). For "science" one might as well read "scholarship" in general, for the ensuing half decade saw the mobilization of scholars of all shades, physicists and humanists alike. For example, the National Defense Research Committee (NDRC) coordinated the contracting of federal research in the development of equipment and war techniques by university physicists, chemists, engineers, and mathematicians.[27] The Manhattan Project, developing the atomic bomb at Los Alamos, drew on physicists including Oppenheimer, Fermi, Teller, Wigner, and von Neumann. Nor was it all "hard science": without a centralized agency to collect and evaluate foreign intelligence, "the inadequacy of the American intelligence apparatus had become conspicuous and critical" (Katz 1989, 2). The response was the Office of Strategic Services (OSS), which drew on academic economists, historians, and sociologists and ultimately became the CIA.

The wartime mathematical activity of interest to us fell under the control of the NDRC. It was established in June 1940 by order of President Roosevelt. Its function was to "correlate governmental and civil research in the fields of military importance outside of aeronautics" and it was chaired by Vannevar Bush (Baxter 1946, 15).[28] The NDRC had five internal divisions catering, respectively, to perceived needs in armor and ordnance; bombs, fuels, and chemicals; communication and transportation; detection, controls, and instruments; patents and inventions. These proceeded by administering contracts with universities and other institutions for defense research. Because of the intense activity and the attendant sense of urgency, these university departments experienced considerable upheaval in the process. For example, the Radiation Laboratory at MIT at its peak employed over four thousand people from all over the country. Its key contribution was the development of shorter-wave radar which yielded better resolution on the radar screen. Other large laboratories, developing underwater sound, were operated in Cali-

27. Baxter 1946 is the official history of the NDRC. Its coverage of the work of the Applied Mathematics Panel is, however, negligible.

28. Aeronautics was in the domain of the National Advisory Committee for Aeronautics (NACA), established in 1938.

fornia and at Columbia, Harvard, and the Woods Hole Oceanographic Institute.

Under the umbrella of the NDRC, mathematicians were engaged in the application of mathematics and probability toward improving the effectiveness of weapons, either through better use of existing ones or the design of new types. They had shown an early keenness to become involved: the American Mathematical Society and the Mathematics Association of America had, in 1940, jointly appointed a War Preparedness Committee with subcommittees on research, preparation for war, and education for service.[29] The two NDRC groups of interest from the point of view of game theory are the Statistical Research Group (SRG) at Columbia, whose work was used primarily by the Army Air Force, and the Anti-Submarine Warfare Operations Research Group (ASWORG) located in Boston and attached to the Navy. The reader should bear in mind, however, that the bureaucratic divisions that seem so neat on paper were, in practice, much less clear: everybody knew everybody else and, to the extent permitted by secrecy, there was much exchange and interaction.

The Statistical Research Group

In late 1942, the NDRC was completely reorganized and the activities of mathematicians were grouped under the control of the Applied Mathematics Panel (AMP), headed by Warren Weaver, with technical assistant Mina Rees.[30] Policy decisions were guided by a committee including R. Courant, G. C. Evans, T. C. Fry, L. M. Graves, O. Veblen, Marston Morse, Sam Wilks, and Weaver. Courant, having left Göttingen in 1933, had established the Courant Institute at NYU. Veblen and Morse were both colleagues at the Institute for Advanced Study with von Neumann, who also served as advisor to the committee (Rees 1980, 609). The AMP set up contracts with eleven universities including Brown, Berkeley, Columbia, Harvard, NYU, Northwestern, and Princeton. Two broad categories of work can be identified among the projects undertaken: first,

29. Among the consultants to the research group were von Neumann (ballistics), Norbert Wiener (computation), and Samuel S. Wilks (probability and statistics) (see Morse 1941, 296).

30. Rees (Ph.D. Mathematics, Chicago 1931) taught at Hunter College until 1943. After the dissolution of the AMP in 1945, she became head of the Mathematics Branch of the Office for Naval Research, which subsequently supported much mathematical activity, including game theory, in universities all over the United States (see "Mina Rees" in Albers 1985).

fluid mechanics, classical dynamics, mechanics of deformable media, and air warfare; second, probability and statistics.

In the first category, the Applied Mathematics Groups operated at NYU, Brown, and Columbia. At NYU, a group under Courant worked on gas dynamics and, in particular, on the theory of explosions both in the air and underwater. This resulted in the *Shock Wave Manual* (1944) and its successor, *Supersonic Flow and Shock Waves* (1948). At Brown, the focus was classical dynamics and the mechanics of deformable media, under W. Praeger. The Columbia group was concerned with aerial warfare, in particular air-to-air gunnery, a departure from the classical applied mathematics of the first two above (Rees 1980, 612). Here the concerns included aeroballistics (the motion of a projectile from an airborne gun), the design of different types of weapon sights, and pursuit curve theory.

The principal work in the second category, probability and statistics, was done by Statistical Research Groups operating at Columbia, Princeton, and Berkeley. The first, the largest and most important of these, was run by Allen Wallis with Harold Hotelling as principal investigator.[31] Samuel Wilks, at Princeton, headed the second group, while Jerzy Neyman at Berkeley ran the third. Wallis's group occupied a portion of a building at West 118 Street, New York, which also had as tenants Columbia's Applied Mathematics Group, mentioned above, and the Strategic Bombing Section of Wilks's Princeton group, run by John D. Williams. Both Williams and Wilks, it appears, interacted very closely with Wallis. During this time, Wallis and Hotelling gathered around them an exceptionally capable group of mathematicians and statisticians including Abraham Wald, J. Wolfowitz, Milton Friedman, James Savage, Abraham Girschick, Frederick Mosteller, and George Stigler (see Wallis 1980). Wald had left Vienna in 1938 with an invitation to the Cowles Commission arranged by Oskar Morgenstern, and then moved to teach at Columbia (Freeman 1968). Wolfowitz, Friedman, and Girschick had all been Hotelling's students during the 1930s.[32] Savage, after

31. At the time that Weaver was still head of the Fire Control division of the NDRC, in 1941, Wallis had left Stanford for the Office of Price Administration (OPA) in Washington (Wallis 1980). At Hotelling's suggestion, Weaver approached Wallis in mid-1942, suggesting that he lead the Columbia SRG, which he did from then until its dissolution in 1945.

32. Girschick, in fact, only spent 1944–45 at the SRG, the remainder of 1939–46 being spent as principal statistician for the Bureau of Agricultural Economics. Affected by his work with Wald, however, after a brief stay at the Bureau of the Census, he went to RAND soon after its creation in 1946 (see *International Encyclopaedia of Statistics*, 398–99).

receiving a Ph.D. from Michigan in 1941, had gone first to the IAS as von Neumann's assistant and then to Cornell and Brown before working for Weaver (Wallis 1980).

As with their Applied Mathematics neighbors, the study of aerial combat was central to the work of the Columbia SRG. It is impossible to adequately describe the 572 reports, memoranda, and substantive letters that resulted, but a sample will give some idea. An early study on alternative ways of placing machine guns on a fighter aircraft involved studying the geometry and tactics of aerial combat. This, in turn, led to work on antiaircraft weapons, aircraft turret sights, and dispersion of aircraft machine guns. A second broad area was the design of the optimal lead angles of aircraft torpedo salvos. This involved the interpretation of photographs of Japanese destroyers to glean information on speed and turning radius. The third field concerned the development of inspection and testing procedures. Out of discussions on the sampling of equipment came the idea that perhaps testing could be stopped before the prescribed sample had all been used, if the information gained thus far somehow suggested that adequate testing had been done. The ratification and formalization of this idea became sequential analysis, developed by Wald.[33] Although Wald himself had already used the minimax idea in his 1939 paper on statistical decision theory, game theory, it appears, was not used as an analytical tool by the group.[34]

One area where the mathematics of games *did* find some use was at ASWORG. Headed by Philip Morse,[35] physicist at MIT, they performed, as their name suggests, operations research directed toward antisubmarine warfare (see Morse 1948, 1951).[36] Morse's group was constituted primarily in response to the presence of German U-boats along the north Atlantic convoy shipping routes. Analyzing data sent in from widespread naval bases, the group made recommendations about the use of equip-

33. The Quartermaster Corps of the Navy apparently made significant economies in inspection using these procedures (see Wallis 1980).

34. Savage, too, was familiar with games, to some extent: "[At lunch one day], Wald discussed some of his ideas on decision theory and Savage . . . remarked that he knew a rather obscure paper that would interest Wald, namely, VonNeumann's 1928 paper on games. Wald laughed and said that some of his ideas were based on that paper" (Wallis 1980, 334).

35. Morse had already done some research work for the Radiation Laboratory and for the Army Air Force (see Tidman 1984, 34).

36. Morse's discussion appears in two sources. Morse 1948 is his Josiah Willard Gibbs lecture to the American Mathematical Society. Morse and Kimball 1951 is the declassified version of a report by the same authors written in 1946 (see Tidman 1984, 102–3).

ment in the pursuit of submarines. For example, taking account of the detection range of equipment and the rate at which a plane or ship could patrol a given area, they calculated a "sighting likelihood curve," which related the probability of sighting to range, visibility, and other factors. Simple probability theory was used to recommend the best way, given the limited resources, to carry out radar and sonar searches of areas of the sea.

ASWORG used game-theoretic analysis in two applications. The first of these was the barrier patrol in which a narrow seaway is patrolled regularly by aircraft in order to prevent the passage of submarines. For example, the Straits of Gibraltar were monitored to keep U-boats out of the Mediterranean, and the area between Brazil and Ascension Island was patrolled in order to catch German ships returning with tin and rubber from Japan and Malaya. The passing submarine, incapable of remaining submerged during its trip through the straits, must choose a point at which to surface. The patrolling airplane must choose a point at which to cross the sound in surveillance. The latter wishes to make contact while the former wishes to avoid it. In order not to have its move predicted, a mixed strategy is preferred; otherwise, the plane would simply choose the point at which it knew the U-boat was going to surface, or the submarine would simply avoid surfacing at the point at which it knew the plane was going to cross. The game is a continuous one: each player must choose a point on the strait of fixed, finite length and does so by applying a probability distribution to the spectrum of possible points. Morse shows how to find the minimax solution to the problem (see 1948, 613–19; 1951, 105–9). When applied, this solution ensures that the lowest probability of contact that the plane can ensure itself is the highest risk of contact that the submarine will have to face. The second application is in the allocation of forces into strategical and tactical components by two opposing armies: Blue and Red.[37] Without going into details, I again note that the minimax solution provides a "safe" option for both sides, given many simplifying assumptions about the relative effectiveness of opponents' forces against each other's production and each other's armies. Any side's choice of tactical force, relative to strategic force, will depend positively on both the opponent's total forces and the opponent's relative weakness in production (see 1951, 73–77).

37. Manichean mathematics! The reader should have little difficulty identifying who's who in this escapade!

Morse finishes with a rallying call,

> The difficulties of solving such problems are not ones of tedious detail, but often due to lack of fundamental techniques. Much more basic research must be carried out before many problems of practical importance can be solved. (1948, 619)

> The studies of von Neumann and Morgenstern show that there are solutions to each problem and show the general nature of these solutions. They do not show, however, the technique for obtaining a solution. . . . A great deal more work needs to be done in finding solutions to various examples before we can say that we know the subject thoroughly. . . . It is to be hoped that further mathematical research can be undertaken on this interesting and fruitful subject. (1951, 109)

The alert reader will observe that the above is largely an unwitting "resurrection" of the prewar work of both Borel and Ville. The Blue and Red force allocation problem is essentially the sort of application suggested, but not analyzed, by Borel in 1938. The barrier patrol is simply a game on the unit square, the solution of which is first offered by Ville, as discussed above. Von Neumann, of course, was familiar with all of this, having been shown it by Morgenstern in 1941 while they wrote their book (see above). He was also close to the applied mathematics being done during the war.[38] He was also completely capable, however, of independently seeing applications and finding solutions to problems presented by Morse and others.

As Morse probably knew, his 1948 talk simply endorsed an agenda for postwar research on games that was already in the making.[39] It was

38. Von Neumann began consulting to governmental organizations after he became a naturalized U.S. citizen in 1937. His first contact was with the Ballistics Research Laboratory (Army Ordnance Department) at Aberdeen, Maryland, to which he was probably introduced by Oswald Veblen. From September 1941 to September 1942 he was attached to Division 8 of the NDRC, working on detonation waves, or how to arrange explosive charges so as to direct and maximize the blast. From September 1942 to July 1943, he advised the Mine Warfare Section of the Navy Bureau of Ordnance on the operations research of mine warfare. This is the work that took him to Washington, D.C., during the last few months of 1942 and to England for the first half of 1943 while working on the *Theory of Games*. In late 1943 and early 1944, he was back at Aberdeen working with Theodore von Karman on aerodynamics and continued to advise the Navy on mines. Beginning in September 1943, he was consultant at Los Alamos on the "Bomb." (See Aspray 1990, 25–27.) This suggests that his work with Morgenstern was a "side interest." From 1943 onward, games would be of interest to von Neumann primarily to the extent that they related to computation.

39. Ed Paxson recalls, "I bought my copy of TGEB [*Theory of Games*] in 1946. I was

in this context of national defense that the theory first gained broader recognition.

Project RAND

As the war ended in 1945, discussions in the War Department, Office for Scientific Research and Development, and branches of the military centered on the prospective return to the campus of the many academics that had been involved in military-oriented research. The links forged by war were about to be dissolved by the peace, and there was a concern that much accumulated experience, along with the possibility for future cooperation, would simply be lost. In March 1946, following a discussion of this type among "Hap" Arnold (Army Air Force Chief of Staff), Dr. Ed Bowles (scientific consultant to the War Department), and some engineers from the Douglas Aircraft Company, Arnold committed $10 million of research funds remaining from the war. Thus Project RAND was born.[40] Initially located under the Douglas roof and then moving to its own quarters in Santa Monica, this group of physicists, mathematicians, and engineers was to conduct a program of research on "intercontinental warfare, other than surface, with the object of advising the Army Air Forces on devices and techniques" (46). In fact the group had more flexibility than this suggests, and Arnold, initially to the disgruntlement of his own military minions, ensured that they were not tied to time or forced to work on immediately applicable ideas. A postwar venue was thus available in which research was easily pursued, was very well paid, and was moreover free of the usual academic duties such as teaching and administration. It was the institutional stabilization of the military-academia symbiosis that had begun during the war.

Reflecting their ostensible purpose, the mathematics group at RAND was labeled the section for the Evaluation of Military Worth (Harris 1990). Its leader was John Williams, who had headed the New York

---

working at the Naval Ordnance Test Station, China Lake, California. Immediately fascinated, I formulated what was later to be called a differential game. This was a real problem, a duel between a destroyer firing at a maneuvering submarine, with allowance for denial of sonar coverage in the destroyer's wake.

"In November of 1946, I spent a day in Princeton with Johnny von Neumann working on this problem. He sketched an approach, and even did some machine coding for his yet unborn machine" (quoted in Shubik 1982, 414).

40. The engineers had advised Bowles in a successful study of the B-29 bomber. Arnold had been supportive of such work and wished to see it continued (see Smith 1966, passim).

branch of Wilks's Princeton Statistical Research Group. Among the others to arrive early were Edwin Paxson (see note 39); Morris "Abe" Girschick, who had also been at the Columbia SRG; Olaf Helmer, who had worked for the AMP in New York; Melvin Dresher, who came from the Office of Price Administration, where Wallis had been before the war; and J. C. C. McKinsey, another mathematician trained at Columbia in the 1930s. Girschick, Helmer, and Dresher were also émigrés: from Russia (1932), Germany (1936), and Poland (1932) respectively (see Kaplan 1983). Exerting a strong influence from a distance were von Neumann, Weaver, and Wilks, all of whom became RAND consultants.[41] Wilks, back at Princeton, directed his best Ph.D. students, among them Ted Harris, toward Williams. Samuel Karlin arrived at the Cal Tech mathematics department from Princeton in 1947 and became consultant to RAND the following summer at the suggestion of his chairman, Henry Bohnenblust, himself a RAND consultant and veteran of wartime operations research in England. Present too were Lloyd Shapley, then with only a B.A. from Harvard and, part-time, David Blackwell of Howard University.[42]

RAND was in this period a center of intellectual ferment. In addition to a constant flow of consultants, lengthy summer sessions were held, to which anyone of note or ability who might have something to contribute was invited.[43] Until the mid-1950s, Santa Monica was the point of reference for those working on matters related to game theory at Princeton, Michigan, and in other military-sponsored research institu-

41. At $200 per month for several hours' work, they were well compensated. Von Neumann in particular was much sought after by many groups: it wasn't flattery when Williams told him, "Paxson, Helmer and all the rest are going to be especially nice to me if I succeed in getting you on the team" (Letter 16 December 1947, VNP, Container 15, File RAND Corp. Contract Corresp.).

42. Karlin would later persuade the exceptionally talented Shapley to return to Princeton for a Ph.D. in mathematics, which he did toward the end of the 1940s (Karlin 1990). Blackwell had met Girschick in 1945 in Washington, D.C., where the latter had lectured on Wald's sequential analysis. There began a collaboration which resulted in, *inter alia*, *Theory of Games and Statistical Decisions*.

43. Beginning in 1948, months-long gatherings were held for the discussion of games and related matters. Those participating were either RAND staff, regular consultants, or visitors from outside. The 1951 guest list, for example, included M. Abramovitz, K. Arrow, E. Domar, J. Duesenberry, J. Marschak, O. Morgenstern, P. Morse, P. Samuelson, and R. Solow. These convocations were generally of the "Secret," if not "Top Secret," variety, and Air Force clearance was required for attendance (for example, see letter of K. E. Wells to O. Morgenstern, 15 June 1950, OMPD, Box 14, File RAND).

tions. This decade saw the "stabilization" of the mathematics of games through the elaboration of its links with other areas in mathematics, with linear programming, and with statistics: the theory became one strut in the framework of ideas developed under immediate postwar military patronage. The concentration on two-person games reflected its relation to military conflict as revealed during the war.[44] As I discuss below, it also resonated with ideas in activity analysis (linear programming) and statistics, and with von Neumann's new interest in computation, all of which were eagerly sponsored by the military. I conclude part 2 by briefly illustrating this stabilization process. This is done to convey to the reader a sense of the broad agenda in which game theory became central; the finer detail must await further treatment.

The Stabilization of Games

Beginning in 1946 with Loomis's completely algebraic proof of the minimax theorem, there issued forth a stream of further proofs, largely falling into two categories. The first of these rest on fixed point theorems, or iterative procedures, and the second on the theory of convex sets (see Kuhn 1952, 71ff.; Luce and Raiffa 1957, Appendix 2). Included in the former are Weyl (1950) and Gale, Kuhn, and Tucker (1950), while the latter includes Nash's (1950) proof of the existence of an equilibrium point for all *n*-person games, of which the minimax is a particular case.[45] Other work utilized the geometric properties of the minimax solution to derive inductive proofs (see Luce and Raiffa 1957, Appendixes 3 and 4). In Dresher et al. 1948, the first collective publication by the RAND mathematics group, the issue of the actual calculation of solutions by matrix methods is raised, and the properties of infinite games under various assumptions about the convexity and continuity of the payoff function are presented. The issue of calculating solutions was particularly important for von Neumann, given his new interest in computation, discussed below. The exploration of continuous games with

44. Mirowski (1991) argues that all of game theory in the immediate postwar period reflected the military influence. Making this case persuasively necessitates showing how all related pursuits such as linear programming, statistical decision theory, experimental games, and computation were part of the military design. One should also show how the work on cooperative games, in Nash 1950 and Thrall et al. 1954, was more the exception than the rule.

45. For a good discussion of various proofs, the reader is directed to Kuhn 1952, his lecture notes on games.

various types of payoff functions was related to hypothetical duels of the bomber-fighter type, and so held the promise of future military value. They also constituted, however, mathematically intriguing problems and offered endless theorem-solving possibilities to the period's most capable mathematicians.[46]

Directly related to the problem of proving the existence of an equilibrium is that of finding, or calculating, the solution. In Kuhn and Tucker (1950) this had become "the principal outstanding problem of zero-sum two-person games" (viii). Although it was not easy for the busy von Neumann to travel to Santa Monica,[47] he was particularly interested in the work linking games and computation being done there. Writing to Warren Weaver in 1948, he said,

> I was very glad to see your comments on RAND's work on the theory of two-person games. I have seen several of their reports, and I need not tell you that I am also very much interested in the fact that some of their attention is now going to this subject. . . . I have spent a good deal of time lately on trying to find numerical methods for determining "optimum strategies" for two-person games. I would like to get such methods which are usable on an electronic machine of the variety which we are planning, and I think that the procedures that I contemplate will work for games up to a few hundred strategies. (1 March 1948, VNP Box 32, File Correspondence W. Weaver)[48]

Indeed, it could well be argued that he remained actively interested in games only to the extent that they related to his work on computation: of his four subsequent papers on the subject, one is concerned with numerical methods for calculating the optimal strategy (1954) and another (Brown and von Neumann) offers a constructive proof intended for "utilization when actually computing the solutions of specific games"(73).[49]

46. The question of what "really motivated" those working on game theory during this period recurs periodically. Personally, I believe motives were mixed. Dresher's declassified version of an earlier RAND volume (1961) is virtually exclusively devoted to games capable, in principle, of military application. Williams (1954) is also particularly concerned with the value of games in strategic considerations. Others, however, such as McKinsey (1952) and Shapley, in general, appear much less "gung-ho" and concerned above all with the mathematics.

47. In October 1951, von Neumann's consulting fee was doubled so that he might pay more attention to RAND than he had hitherto done. (See letter from Alex Mood [RAND] to von Neumann, 1 October 1951, VNP Box 15, Folder RAND.)

48. For various requests for advice from McKinsey and Paxson to von Neumann, see Letters, VNP, Box 25, File RAND Corporation.

49. For an exploration of von Neumann's work in computation see Aspray 1990. In 1943, while on a research trip to Britain for the Navy, von Neumann wrote to Veblen of his newly

The theoretical links between games and other areas such as statistics and linear programming were soon established. Following Wald 1945, which showed the application of the minimax theorem in statistical decision theory, the decision process being characterized as a game against Nature, these ideas were further developed in Arrow, Blackwell, and Girschick 1949 and Blackwell and Girschick 1954. In Koopmans 1951, both Dantzig and Gale, Kuhn, and Tucker demonstrate the equivalence of the tasks of solving a game and a linear programming problem, thereby drawing on activity analysis to provide *constructive* existence proofs in game theory. At RAND, 1952 also saw the first work in experimentation on games (see Flood 1958). With regard to the question of military application, both Haywood 1951 and Caywood and Thomas 1955 offer simple, stylized examples. These illustrate better the *confidence* the military held in the mathematics as a tool rather than the direct *usefulness* of the ideas in this context. Finally, out of this milieu in the early 1950s came the first textbook on game theory, McKinsey 1952, and the first popularizations, McDonald 1950 and Williams 1954. While the latter books became immediately popular, McKinsey's book was perhaps a little too austere for the economics readership of the 1950s and was supplanted several years later by the more "user-friendly" Luce and Raiffa 1957.[50]

Conclusion

This paper supports a relatively simple claim about the evolution of what we now know as game theory. For the thirty-five years preceding the end of World War II, the thinking on games was limited, disparate, and disconnected. In the absence of anything that could be regarded as discourse, the "theory," if it could be called such, was of limited meaning to a limited number of people. What ultimately gave it life was not Morgenstern's hope for its transformation of economics, but the changed postwar environment for mathematical research, the strong military interest in the theory, and the attention given to it by the mathematicians under their patronage. The confluence of these various streams created the context in which the theory was first stabilized. There re-

developed "obscene interest in computational techniques" (quoted in Aspray 1990, 27). That game theory could be sustained despite his diminished interest was one sign of its maturation as a field of applied mathematics.

50. I owe this observation about the early textbooks to William Riker.

mains to be written a historical account of the passage of game theory out of military research into various universities with the encouragement of the ONR, its assimilation into political and biological science, and, above all, the halting process by which it gradually transformed the canon in microeconomics.

References

Albers, Donald J., and G. L. Alexanderson, eds. 1985. *Mathematical People*. Boston: Birkhauser.

Arrow, Kenneth. 1989. *John von Neumann and Modern Economics*. Oxford: Clarendon.

Arrow, Kenneth, D. Blackwell, and M. A. Girschick. 1949. Bayes and Minimax Solutions of Sequential Decision Problems. *Econometrica* 17 (July-October): 213–44.

Aspray, William. 1990. *John von Neumann and the Origins of Modern Computing*. Cambridge, Mass.: MIT Press.

Baxter, James Phinney. 1946. *Scientists against Time*. Boston: Little, Brown.

Blackwell, D., and M. A. Girschick. 1954. *Theory of Games and Statistical Decisions*. New York: Wiley.

Borel, Émile. 1914. *Le Hazard*. Paris: Alcan.

———. [1921] 1953. La théorie du jeu et les équations intégrales à noyau symétrique. *Comptes Rendus Hebdomadaire de l'Académie des Sciences*. 173.25 (19 December): 1304–8. Translated by L. J. Savage as The Theory of Play and Integral Equations with Skew Symmetric Kernels. *Econometrica* 21 (January): 97–100.

———. 1923. Sur les jeux où interviennent l'hasard et l'habileté des joueurs. *Association Française pour l'Avancement des Sciences*.

———. [1924] 1953. Sur les jeux où interviennent l'hasard et l'habileté des joueurs. *Théorie des Probabilités*. Paris: Librairie Scientifique, Hermann. Translated by L. J. Savage as On Games that Involve Chance and the Skill of Players. *Econometrica* 21 (January): 101–15.

———. 1926. Un théorème sur les systèmes de formes linéaires à déterminant symétrique gauche. *Comptes Rendus de l'Académie des Sciences* 183.21 (22 November): 925–27, avec erratum, 996.

———. [1927] 1953. Sur les systèmes de formes linéaires à déterminants symétrique gauche et la théorie générale du jeu. *Comptes Rendus de l'Académie des Sciences* 184.2 (10 January): 52–54. Translated by L. J. Savage as On Systems of Linear Forms of Skew Symmetric Determinants and the General Theory of Play. *Econometrica* 21 (January): 116–17.

———. [1950] 1965. *Elements of the Theory of Probability*, translated by John E. Freund. Englewood Cliffs, N.J.: Prentice-Hall.

Borel, Émile, and L. Painlevé. 1910. *L'Aviation*. Paris: Alcan.

Borel, Émile, et al. 1938. *Traité du calcul des probabilités et de ses applications*. Tome 4, Fasc. 2, Applications aux Jeux de Hasard. Paris: Gauthier-Villars.

Brown, G. W., and J. von Neumann. 1950. Solutions of Games by Differential Equations. In *Contributions to the Theory of Games*, vol. 1, edited by H. Kuhn and A. W. Tucker. Princeton: Princeton University Press.

Caywood, T. E., and C. J. Thomas. 1955. Applications of Games Theory in Fighter versus Bomber Combat. *Journal of the Operations Research Society of America* 3.4:402–11.

Collingwood, E. F. 1959. Émile Borel. *Journal of the London Mathematical Society* 34.4 (October): 488–512.

De Possel, René. 1936. *Sur la théorie mathématique de jeux de hasard et de reflexion*. Paris: Hermann & Cie. (Conférences du Centre Universitaire Méditerranéen de Nice, publiée sous la direction de M. Paul Valéry. Actualités Scientifiques et Industrielles, N. 436.)

Dore, Mohammed, et al., eds. 1989. *John von Neumann and Modern Economics*. Oxford: Clarendon.

Dresher, Melvin. 1961. *Games of Strategy*. Englewood Cliffs, N.J.: Prentice-Hall.

Dresher, Melvin, et al. 1948. Mathematical Theory of Zero-Sum Two-Person Games with a Finite Number or a Continuum of Strategies. RAND Corp. September 3.

Fermi, Laura. 1968. *Illustrious Immigrants*. Chicago: University of Chicago Press.

Fisher, Franklin. 1989. Games Economists Play: A Noncooperative View. *RAND Journal of Economics* 20.1:113–24.

Flood, Merrill. 1958. Some Experimental Games. *Management Science* 5.1 (October): 5–26.

Fréchet, Maurice. 1953. Émile Borel, Initiator of the Theory of Psychological Games and Its Application. *Econometrica* 21 (January): 118–27.

———. 1955. *Les mathématiques et le concret*. Paris: Presses Universitaires.

———. 1965. La vie et l'oeuvre d'Émile Borel. *L'Enseignement Mathématique*. 1.1:1–97.

Freeman, Harold. 1968. Wald, Abraham. *International Encyclopedia of the Social Sciences*, vol. 16. New York: Macmillan.

Gale, D., H. Kuhn, and A. Tucker. 1950. On Symmetric Games. In *Contributions to the Theory of Games*, vol. 1, edited by H. Kuhn and A. W. Tucker. Princeton: Princeton University Press.

Goldstine, Herman. 1972. *The Computer from Pascal to von Neumann*. Princeton: Princeton University Press.

Harris, Ted. 1990. Interview, 29 February. Los Angeles.

Haywood, Col. Oliver G. 1951. Military Doctrine and the von Neumann Theory of Games. *RAND RM-528*, February.

Heims, Steve J. 1980. *John von Neumann and Norbert Wiener*. Cambridge, Mass.: MIT Press.

Ingrao, Bruna, and Giorgio Israel. 1990. *The Invisible Hand*. Cambridge, Mass.: MIT Press.

International Encyclopedia of Statistics. 1978. New York: Free Press.
Jacquemin, Alex. 1987. *The New Industrial Organization*. Cambridge, Mass.: MIT Press.
Kac, Mark. 1985. *Enigmas of Chance*. New York: Harper & Row.
Kaplan, Fred. 1983. *The Wizards of Armageddon*. New York: Simon & Schuster.
Karlin, Sam. 1990. Interview, 2 March. Palo Alto.
Katz, Barry. 1989. *Foreign Intelligence*. Cambridge, Mass.: Harvard University Press.
Koopmans, Tjalling C., ed. 1951. *Activity Analysis of Production and Allocation*. New York: Wiley.
Kreps, David. 1990. A Course in Microeconomic Theory. Princeton: Princeton University Press.
Kuhn, H. 1952. *Lectures on the Theory of Games*, issued as a report of the Logistics Research Project, Office of Naval Research. Princeton: Princeton University Press.
Kuhn, H., and A. W. Tucker, eds. 1950. *Contributions to the Theory of Games*, vol 1. Princeton: Princeton University Press.
Kuratowski, Kazimierz. 1980. *A Half-Century of Polish Mathematics*. Oxford: Pergamon.
Loomis, Lynn H. 1946. On a Theorem of von Neumann. In *Proceedings of the National Academy of Science* 32.8 (August): 213–15.
Luce, R. Duncan, and H. Raiffa. 1957. *Games and Decisions*. New York: McGraw-Hill.
McDonald, John. 1950. *Strategy in Poker, Business, and War*. New York: Norton.
McKinsey, J. C. C. 1952. *Introduction to the Theory of Games*. New York: McGraw-Hill.
Mensch, A., ed. 1966. *Theory of Games: Techniques and Applications*. New York: American Elsevier.
Mirowski, Philip. 1991. When Games Grow Deadly Serious: The Military Influence on the Evolution of Game Theory. In *Economics and National Security*, edited by Craufurd D. Goodwin. Durham, N.C.: Duke University Press.
Morgenstern Papers: Diaries, Oskar. Duke University Library, Durham, N.C. (OMPD).
Morgenstern Papers, Oskar. Duke University Library, Durham, N.C. (OMP).
Morgenstern, Oskar. 1928. Wirtschaftsprognose. Vienna: Julius Springer.
———. [1935a] 1976. The Time Moment in Economic Theory. *Zeitschrift für Nationalökonomie* 5.5:433–58. (Translation in Schotter 1976.)
———. [1935b] 1976. Perfect Foresight and Economic Equilibrium. *Zeitschrift für Nationalökonomie* 6.3:337–57. (Translation in Schotter 1976.)
———. [1936] 1976. Logistics and the Social Sciences. *Zeitschrift für Nationalökonomie* 7.1:1–24. (Translation in Schotter 1976.)
———. 1941a. Quantitative Implications of Maxims of Behavior. Oskar Morgenstern Papers, Duke University Library, Box 49, file Maxims of Behavior 1939–70.
———. 1941b. Professor Hicks on Value and Capital. *Journal of Political Economy* 49.3 (June): 361–93.

———. 1976. The Collaboration between Oskar Morgenstern and John von Neumann on the Theory of Games. *Journal of Economic Literature* 14.3 (September): 805–16.

Morse, P. M. 1948. Mathematical Problems in Operations Research. *Bulletin of the American Mathematical Society* 54.7 (July): 602–21.

Morse, P. M., and George E. Kimball. 1951. *Methods of Operations Research*. New York: Technology Press and Wiley. (Originally in classified form as 1946, same title, OEG Report 54.)

Morse, P. M., and William L. Hart. 1941. Mathematics in the Defense Program. *American Mathematical Monthly* 48 (May): 293–302.

Nash, Jr., John. 1950. Equilibrium Points in N-Person Games. *Proceedings of the National Academy of Sciences*. 36.1 (January): 48–49.

Rees, Mina. 1980. The Mathematical Sciences and World War II. *American Mathematical Monthly* 87.8 (October): 607–21.

Reid, Constance. 1970. *Hilbert*. New York: Springer.

Rellstab, Urs. 1991. From German Romanticism to Game Theory: I. Oskar Morgenstern's Vienna in the 1920s. Photocopy, Department of Economics, Duke University.

———. 1992. New Insights into the Collaboration between John von Neumann and Oskar Morgenstern on the *Theory of Games and Economic Behavior*. *HOPE*, this issue.

Rives, Jr., Norfleet W. 1975. On the History of the Mathematical Theory of Games. *HOPE* 7.4:549–65.

Schotter, Andrew, ed. 1976. *Selected Economic Writings of Oskar Morgenstern*. New York: NYU Press.

Science News Letter. 1937. Princeton Scientist Analyzes Gambling: You Can't Win. April 3.

Shubik, Martin. 1982. *Game Theory in the Social Sciences*. Cambridge, Mass.: MIT Press.

Smith, Bruce. 1966. *The RAND Corporation*. Cambridge, Mass.: Harvard University Press.

Steinhaus, Hugo. 1960. Definition for a Theory of Games and Pursuit. *Naval Research Logistics Quarterly* 7.2:105–8.

Taub, Alfred H., ed. 1963. *John von Neumann: Collected Works*, 4 vols. New York: Macmillan.

Thrall, Robert, C. H. Coombs, and R. L. Davis. 1954. *Decision Processes*. New York: Wiley.

Tidman, Keith R. 1984. *The Operations Evaluation Group*. Annapolis: Naval Institute Press.

Tirole, Jean. 1988. *The Theory of Industrial Organization*. Cambridge, Mass.: MIT Press.

Tucker, A. W., and R. D. Luce, eds. 1959. *Contributions to the Theory of Games*, vol. 4. Princeton: Princeton University Press.

Ville, Jean. 1938. Sur la théorie génerale des jeux où intervient l'habileté des joueurs. In Borel et al. 1938.

von Neumann Papers, Library of Congress, Washington, D.C. (VNP).
von Neumann, John. 1927a. Mathematische Begrundung der Quantenmechanik. *Nachrichten von der Gesellschaft der Wissenschaften Su Göttingen*. (See also Taub 1963, vol. 1.)
———. 1927b. Wahrscheinlichkeitstheoretischer Aufbau der Quantenmechanik. *Nachrichten von der Gesellschaft der Wissenschaften Su Göttingen*. (See also Taub 1963, vol. 1.)
———. 1927c. Thermodynamik Quantenmechanischer Gesamtheiten. *Nachrichten von der Gesellschaft der Wissenschaften Su Göttingen* (See also Taub 1963, vol. 1.)
———. 1928a. Sur la théorie des jeux. *Comptes Rendus de l'Académie des Sciences*, vol. 186.25 (18 June): 1689–91.
———. 1928b. Zur Theorie der Gesellschaftsspiele. *Mathematische Annalen* 100: 295–320. (Translation by S. Bargmann in Tucker and Luce 1959.)
———. 1929a. Allgemeine Eigenwerttheorie Hermitescher Funktionaloperatoren. *Mathematische Annalen* 102:49–131.
———. 1929b. Zur algebra der Funktionaloperatoren und Theorie der normalen Operatoren. *Mathematische Annalen* 102:370–427.
———. 1929c. Zur Theorie der unbeschrankten Matrizen. *J. F. Math*. 161:208–36.
———. 1940. *Theory of Games I*, manuscript in Oskar Morgenstern Papers, Duke University Library, Box 51, File: John von Neumann, 1940–48.
———. 1941. *Theory of Games II*, manuscript in Oskar Morgenstern Papers, Duke University Library, Box 51, File: John von Neumann, 1940–88.
———. 1954. A Numerical Method to Determine Optimum Strategy. *Naval Research Logistics Quarterly* 1:109–15.
———. [1932] 1955. *Mathematical Foundations of Quantum Mechanics*, translated by Robert Beyer. Princeton: Princeton University Press.
von Neumann, John, and Oskar Morgenstern. 1944. *The Theory of Games and Economic Behavior*. Princeton: Princeton University Press.
Wald, A. 1945. Statistical Functions which Minimize the Maximum Risk. *Annals of Mathematics* 46.2 (April):265–80.
Wallis, W. Allen. 1980. The Statistical Research Group. *Journal of the American Statistical Association* 75.370:320–30, and Rejoinder 75.370:334–35.
Weintraub, E. Roy. 1985. *General Equilibrium Analysis*. Studies in Appraisal. Cambridge: Cambridge University Press.
Weyl, Hermann. 1950. Elementary Proof of a Minimax Theorem Due to von Neumann. In Kuhn and Tucker 1950.
Williams, John D. [1954] 1966. *The Compleat Strategyst*. New York: McGraw-Hill.

New Insights into the Collaboration between John von Neumann and Oskar Morgenstern on the *Theory of Games and Economic Behavior*

Urs Rellstab

Introduction

Take as given the mathematical genius of John von Neumann and his early contribution to game theory. Assume that Oskar Morgenstern did not himself contribute much to the further development of the theory of games after their collaboration. The question then must be asked, "What was Morgenstern's contribution to the theory of games?" The asymmetry between von Neumann and Morgenstern in their collaboration on the theory of games is not hidden. Their names are not in alphabetical order on the cover of the *Theory of Games and Economic Behavior* 1944 (hereafter TGEB); and the preface states that "the theory had been developed by one of them since 1928" (von Neumann and Morgenstern 1953, v).

Still, what really happened intellectually between the two men remains hidden. And if the relationship was asymmetrical, what was the nature of Morgenstern's contribution to the TGEB? Did it go beyond his Sherlock Holmes story and the discovery of Ville's proof? Some scholars of Morgenstern remain vague on this point. Shubik tells the story of a cocktail party where someone asks: "What has been Morgenstern's contribution to mathematical economics?" The immediate reply is "Abraham Wald and John von Neumann." Shubik comments: "If that had been Morgenstern's only contribution to economics, in my estimation that would have been enough and would have been a contribution considerably larger than many of his peers. . . . It was Morgenstern's ability both to see the relevance of the work of these two great mathema-

I would like to thank the Swiss National Science Foundation for financial support and the following for comments on an earlier draft: Robert Leonard, Neil de Marchi, Philip Mirowski, Bob Coats, Mary Morgan, and E. Roy Weintraub. I also want to thank Kirsten Fisher for checking my translations from Oskar Morgenstern's diaries as well as Anne-Marie Rasmussen and Forrest Smith for editing the article.

ticians and to persist in getting them to work on problems of economic significance" (1981, 9). Schotter suggests that he could not think of another economist at the time "who could have walked into a room with John von Neumann and walked out later with a 600-page book on the theory of games complete with economic examples. That fact speaks for itself" (1992). Actually, the facts are not very "eloquent," but are rather silent; the evidence is diffuse.

From the moment the draft of TGEB was finished, Morgenstern was well aware that he would owe some explanation to the public about his contribution. He promised, "I have to write something about the history of the book (and about my minimal share in it; it seems however that my effect on Johnny was like some kind of a catalytic factor)" (OMPD, Box 13, 1 January 1943). In 1976 he partly fulfilled his promise (Morgenstern 1976a). Again, Morgenstern was aware of many questions. "A great deal would have to be said about what actually went on intellectually between the two of us" (814). In fact, as Morgenstern's diaries reveal, there was not much of an intellectual struggle between the two. Morgenstern, as "catalytic factor," was mainly just a spectator to the reactions that occurred in von Neumann's mind. Of course, his questions, his enthusiasm, and in particular his indefatigable work as secretary sped up the process, but it can be shown that some of the issues that were important to him (and on which he worked) did not enter the book.[1] Still, Morgenstern wanted to say more about the collaboration and he fed the reader with hopes. "A fuller account with precise dates may follow some other time" (805). Unfortunately, he was not able to give this fuller account; cancer overtook him and he died a year later. But there is another source that can close the gap. Morgenstern was quite zealous in keeping his diaries. A close look at the diaries and the Morgenstern papers reveals

1. In addition, I think that the static nature of the TGEB contradicts Morgenstern's earlier writings. From *Wirtschaftsprognose* to his writings in the thirties, one of his main concerns was the inclusion of the element of time. Why did he give up this important issue? The paragraph about statics and dynamics in the TGEB—written by Morgenstern—sounds very much like an excuse (von Neumann and Morgenstern 1953, 44–45). Part of the story might be that Morgenstern became the victim of his own example of Sherlock Holmes. The example illustrates in *Wirtschaftsprognose* that even repeated public forecasts cannot include the reactions of the individual economic actors to the forecast. This all occurs "in time": it is a problem of diachronic interdependence. On the other hand, the problem of Sherlock Holmes is one of synchronic interdependence: it can be solved, as Morgenstern comments himself, right on the spot (in the station) or at a single point of time. This hypothesis cannot be further outlined in this article.

additional information about the collaboration between von Neumann and Morgenstern. It can be shown that Morgenstern's official report of the collaboration, published in 1976, does not correspond to the record preserved in his diaries in various respects. This is because his retrospective account represents a truly separate perspective, having passed through the filter of more than thirty years. Even so, there are some facts that should not be affected by the change of perspective, especially since it is likely that Morgenstern himself consulted the diaries in order to refresh his memory.

This article outlines the chronology of the collaboration between von Neumann and Oskar Morgenstern on the TGEB, following the unofficial report in Morgenstern's diaries and papers. I make extensive use of the diaries in the hope that the reader can catch the atmosphere of the collaboration. Basically, the work to be done is to solve what in Austrian terms is the *Zurechnungsproblem* (problem of imputation) between two factors of production: John von Neumann and Oskar Morgenstern.

The most striking result of this research is that von Neumann elaborated his theory of games after his 1928 paper, before the actual collaboration with Morgenstern started. On the other hand there is absolutely no indication that Morgenstern wrote a paper on games (the one promised for the economists; see Morgenstern 1976a, 808) before the two decided to work together in 1941.

Focusing on the technical side of the theoretical innovation that the theory of games represents, it becomes very clear that Morgenstern's share was minimal. However, Morgenstern encouraged von Neumann in discussions and with a paper on "maxims of behavior" to further develop his ideas and make them applicable to the social sciences. He offered von Neumann, in his role as a "catalytic factor," an economic context that challenged neoclassical economics.

Morgenstern's Unofficial Report of the Collaboration

Morgenstern remembered a tea with Niels Bohr at Fine Hall in Princeton at the beginning of 1939 where he met von Neumann for the second time (Morgenstern 1976a, 807).[2] This meeting and "an instantaneous

2. They met for the first time in 1938 in Princeton soon after the semester started, a meeting neither of them remembered later.

meeting of minds" initiated the exchange of papers. A lot of discussion followed, partly because Morgenstern needed von Neumann to explain his 1928 article "Zur Theorie der Gesellschaftsspiele." Morgenstern decided to write a short paper explaining to economists the essence of the theory of games. This, in turn, necessitated further contacts with von Neumann, who asked Morgenstern in the fall of 1940, "Why don't we write this paper together?" (808). Their joint project grew: from a paper to a pamphlet, and from a pamphlet to a book of more than 600 pages. Morgenstern remembers that the work took place mainly during 1941 and 1942. Although von Neumann moved to Washington in 1942, they were able to finish the book by the end of the year.

This is the condensed form of Morgenstern's official version. An important feature of this official narration is that von Neumann did not work in a systematic way on games after his 1928 paper, and that it was Morgenstern who took the next step by writing the paper for economists. But this official account may be incomplete. By the time of the collaboration with von Neumann, Morgenstern was keeping extensive private diaries, and the story in the diaries contradicts the official version in some respects.

Unfortunately however, these diaries contain two gaps. Two periods —Morgenstern's originally planned stay in the United States, from February to May 1938; and from March 1939 to January 1940—are not covered. Morgenstern's decision to stay in the United States after the "Anschluss" of Austria and the steps that led to his employment at Princeton took place during the first gap. In a letter to Bertil Ohlin, Morgenstern writes that during his visit as a Carnegie Professor, he received six invitations from various universities to stay in the United States. He gives two explicit reasons for choosing Princeton, yet no reference to von Neumann's presence there: "For many reasons I choose Princeton, partly because it is but one hour from New York partly, however, because they wanted me already last year which shows that they wanted me independently of the events in Austria. This gives it a very nice psychological side" (Letter to Ohlin, 21 July 1938, OMP, Box 8, Correspondence Chronological 1925–38).

During the second gap, according to the official account, von Neumann and Morgenstern must have met several times for discussions. But according to the later diaries, in 1939 there was no regular contact between the two. So it seems that not much is missing because of this second gap.

Let us proceed step by step, following Morgenstern's unofficial report. Although Morgenstern and von Neumann had known each other since the fall of 1938, their relationship was fairly limited until April 1940, when the two minds really met during the month-long absence of von Neumann's wife, Clari. Von Neumann's presence at Niels Bohr's tea at Fine Hall in Princeton in February 1939 is not mentioned in the diaries (OMPD, 15 February 1939). Those show that Morgenstern was greatly fascinated by Bohr's observations about the disturbance of experiments by the observer, but there is no mention of von Neumann nor a discussion on games. Morgenstern and von Neumann had their first intense discussion in October 1939 (OMPD, 4 April 1940). At this time Morgenstern was taking notes for a paper on maxims of behavior. He wrote at the end of the same month to Frank Knight about his disillusionment with the economics department at Princeton and commented that he got stimulation from mathematicians. I quote at length from his letter.

> I understand that you are writing a review of Hutchison for the Journal of Political Economy, and I shall be very much interested to see what you have to say about this book which I also discussed last year here at Princeton. I did not find very much interest, however, because there are only a few people, if any, interested in methodological questions. Those with whom to discuss such problems are principally the mathematicians, of which we have some excellent ones in town. I have now been stimulated by these talks and proceeded to jot down notes on a further paper of what I called maxims of behaviour. In this paper I shall endeavor to investigate a very curious relationship between the quantitative limits which maxims may have. I hope to be able to show you something of this in the not too distant future, and shall welcome your critisisms very greatly. (Letter to Knight, 14 October 1939; OMP, Box 6, Correspondence: 1928–39, Knight Frank H.)

Apparently the discussion with von Neumann fascinated Morgenstern but the event was isolated and without immediate consequence. At the beginning of 1940, Morgenstern was invited for dinner three times to von Neumann's home (25 January 1940, 9 March 1940, 1 April 1940 [OMPD]). On each of these occasions von Neumann's wife was present, and the conversation evidently did not center on games.

The situation changed dramatically in April 1940, when the prelude to the collaboration ended. However, before the actual collaboration began in June or July 1941, there followed a phase of intense discussions. Morgenstern writes:

> In the evening John von Neumann came and *we* had a discussion for nearly four hours: Maxims of behavior (he sees completely what it is all about and how difficult it is), about games, about basic questions of mathematics, he talked especially about Goedel and general philosophy of science. It has been a long time since I spent such an interesting evening. He stuck totally to the point and we may continue tomorrow. It is a pity that we have not discussed more often together; but I did not think I could contribute to his problems. I should have remembered our first discussion last October. Because he will be alone for four weeks, we can come together more easily. (4 April 1940, OMPD)

Morgenstern's wish was fulfilled. In April 1940, the two men met for several long discussions. It seems that by this time von Neumann had become interested in Morgenstern's work. Von Neumann read Morgenstern's paper "Das Zeitmoment in der Wertlehre," praised it, and recommended that Morgenstern extend it into a book on risk, foresight, and revenue.[3] At the beginning of May, when von Neumann left for a trip to the West Coast, Morgenstern noted: "We spent several evenings together and discussed for many hours. Mainly mathematical economics" (3 May 1940, OMPD). And he continues, "It would be wonderful, if we could write a book together" (3 May 1940, OMPD).

But this was, at that moment, a mere wish. During the summer Morgenstern worked on a paper on "business cycles," which he completed on the twelfth of November (OMPD).[4] In August, Morgenstern and von Neumann met again. Von Neumann had not simply been enjoying the landscape on his trip to the West. In Seattle, he lectured on games and announced to Morgenstern in a letter that he would be able to tell

3. For the English version see Schotter 1976, 151–67, "The Time Element in Value Theory."

4. Several times he mentions this paper in his diaries, referring to it rather vaguely as "Aufsatz" (i.e., OMPD, 21 July 1940, 20 August 1940, 5 September 1940, and 21 September 1940). Morgenstern also talked about the paper with von Neumann. This paper was never published, but it is part of a much larger work on "International Financial Transactions and Business Cycles" (Morgenstern 1959). Morgenstern started this work at the suggestion of Wesley C. Mitchell, and it was published twenty years later, "after countless interruptions," as a study by the National Bureau of Economic Research (viii). One of these interruptions was, of course, the theory of games, "which inserted itself while this investigation was in its early phases."

him some new things about four-or-more-person games (16 July 1940, OMP, Box 51, "subject file von Neumann 1940–1945"). Concerning the discussions on this issue, Morgenstern observed, "These combinatorical things are complicated and not very clear. He [von Neumann] hopes that a stimulation from me will allow an arithmetical treatment, which would make great tools available" (20 August 1940, OMPD). The personal relationship between von Neumann and Morgenstern intensified during August and September 1940. By the end of September the two had put aside German formality and had begun using first names: from then on Morgenstern referred to von Neumann as "Johnny." Von Neumann certainly explained his theory of games in detail to Morgenstern; but in this early stage of general discussion, before the decision to collaborate was made, Morgenstern supplied him with criticism about established economic thinking. "Recently lunch with Johnny. I told him about some criticisms of the theory of indifference curves and he agreed totally with me. He is also convinced that marginal utility is *not* eliminated. I will work all that into the Hicks review. It would be good if he were to write a mathematical supplement" (19 October 1940, OMPD).

On other occasions Morgenstern provided von Neumann with parallels between economic theories and games. He saw clearly that economic problems are not zero-sum games.

> Yesterday I showed him the contract curve; it stands in relation to his games because it makes a difference who goes first. He will think about all of this. Moreover, I showed him the causal relationships in BB's [Böhm-Bawerk's] price theory where, as one can easily see, one obtains different results, depending on the assumed knowledge of the other's position. It differs from a game because a profit is possible. (9 November 1940, OMPD)

This discussion occurred after Morgenstern visited one of von Neumann's lectures on games for the first time. The lectures were not terribly well attended (Morgenstern 1976a). Von Neumann did more than lecture on games. In October 1940, he wrote a paper on games with the title "Theory of Games I, General Foundations" (Box 51, subject file "Neumann, John von, 1940–45" OMP). It deals with the value function and the concepts of domination and solution. This paper was followed by a second in January 1941 entitled, "Theory of Games II, Decomposition Theory." Although these working papers do not name an author, the mathematical style and the fact that Morgenstern put the papers in a folder with von Neumann's other writings indicate that von Neumann was the author.

On the other hand, there is no indication in the diaries that Morgenstern was working on a paper on games yet. If he really had been working on such a paper, it would have been odd not to have mentioned it in the diaries, not only because von Neumann was so important to him, but because he normally wrote much about his work. Thus, it is clear that von Neumann further elaborated his theory of games before the two actually started to collaborate, and Morgenstern did not begin to write prior to the actual collaboration. This can be further illustrated.

On 17 May 1941, von Neumann asked Morgenstern to write a paper on "maxims," referring to "maxims of behavior," a topic that had been Morgenstern's main interest since the beginning of their discussions (see 5 April 1940, OMPD). On the very next day, happy about von Neumann's request, Morgenstern began work on the paper: "I have a mountain of notes (about the subject) and I will write a short paper, maybe 5000 words" (17 May 1941, OMPD). This paper, "the maxims," had a direct impact on von Neumann's decision to collaborate with Morgenstern. The full title is "Quantitative Implications of Maxims of Behavior." The original—dated May 18, 1941, on the cover—can be easily located in the Morgenstern Papers at Duke University (OMP, Box 49, subject file "Maxims of Behavior 1939–1970"). Although it was never published, Morgenstern wanted to publish it later on several occasions, and both von Neumann and, later, Gödel urged him to do so[5] (28 January 1948, OMPD). The paper is particularly valuable because it reflects Morgenstern's pre–game-theoretical concerns. It is written from the point of view of an economist in search of a method to deal with the interdependence of economic behavior. For Morgenstern, the main problem of this interdependence lies in the fact that it limits the economic actor's capacity to foresee the future; but the reader will look in vain for any quantitative aspect. Morgenstern notes, "It cannot be hoped to do more than to pose the problem qualitatively because its full treat-

5. There exists a published paper by Morgenstern (1976b) with a similar title, "Some Thoughts on Maxims of Behavior in a Dynamic Universe." But this paper differs in content from the "maxims" written for von Neumann more than thirty years earlier. In "Some Thoughts on Maxims . . ." the importance of the notion of time reappears and game theory is described as dealing with just one case of restricted maxims of behavior, namely, the analysis of interdependent human actions. Other constraints of behavior are of a technological, legal, or moral nature.

ment would require techniques of analysis which are not available at the present time ("Maxims," 1). And the paper reads more like an invitation to von Neumann to collaborate: "Consequently, it will be necessary to leave it to the professional mathematicians to work out solutions and it is not exaggerated to think that they will have a very hard nut to crack. Reference should be made to the studies on the theory of games by J. von Neumann which were recently extended in a series of lectures" (22). As we know, the invitation was accepted. Morgenstern wrote in his diaries, "The maxims are lying around and need to be worked on. In the meantime I began a paper with Johnny about games, minimax, bilateral monopoly and duopoly" (12 December 1941, OMPD).

This is the first time the diaries mention a paper on games linked to economic theory; this is probably the paper Morgenstern mentioned in 1976.[6] According to the diaries, it was written by Morgenstern with an enormous effort right after von Neumann declared, in June or July 1941, his wish to collaborate.[7] Morgenstern began to write on the basis of von Neumann's manuscript(s). His explicit purpose was to write an introduction to them (12 July 1941, OMPD).

Beside the fact that Morgenstern began to write on games very late, it is not clear why in his official account he put the actual beginning of his collaboration with von Neumann during the fall of 1940 instead of the summer of 1941 (Morgenstern 1976a, 808). In the first draft of his paper on the collaboration, written in August 1971 at Blue Mountain Lake, New York, Morgenstern states that he wrote from memory only (OMP, Box 22, writings and speeches, "The Collaboration between Oskar Morgenstern and John von Neumann on the Theory of Games, 1976"). But it seems that he consulted the diaries before the publication of his account because the essential paragraphs of the diaries are marked with a red felt-tipped pen. In any case, he still missed the correct dates. Did he confuse the dates, or would he have liked the actual collaboration to have started earlier?

At the time the collaboration really started, Morgenstern's joy about his work with von Neumann gave him the hope that their joint work

6. In the article about the collaboration, Morgenstern makes the reader think that this paper on games was written before the actual collaboration with von Neumann started.

7. What follows in the next one-and-a-half years is intellectually the most stimulating and productive time of Morgenstern's life. "Some people tell me, that they have never seen me in such a good form and good mood. This is due to Johnny who completely woke me up. Suddenly wishes are being fulfilled" (2 December 1941, OMPD).

could continue: "It would be nice if in the next few years Johnny and I could write a general economic theory. It should be exact whenever possible and should say what one doesn't know. This is the book I have been dreaming of for years. It should be possible" (30 July 1941, OMPD). From the beginning Morgenstern was aware of the importance of the work with von Neumann, because it touched the fundamentals of the subjective theory of value. Several emotions characterized Morgenstern's entries over the next eighteen months: first, his joy over von Neumann's willingness to collaborate, which was only overshadowed by the fear that he might lose interest; second, his appreciation of the chance to work with one of the greatest living mathematicians and his admiration for the way von Neumann conceptualized problems; third, his recognition of the complicated nature of the problems; and finally, the presentiment that it might be a long time before their theory replaced existing theories, which in his view were no longer accurate.[8]

By the beginning of August, the manuscript amounted to fifty pages, and Morgenstern expected it to grow to one hundred (4 August 1941, OMPD). Yet, this would be much too long for a publication in the *Journal of Political Economy*. Morgenstern portrayed von Neumann's reaction to his manuscript as positive, and he felt that he knew along what lines to proceed. Of course, he was proud of this, and he commented that it was not easy for him to represent von Neumann's theory in a simplified but still correct form.

However, I would argue that by the end of the first two months of the project there already was an asymmetry of effort on the theory of games. It was Morgenstern who did the job in rather loose interaction with von Neumann, who at this time was working on a paper on "Operatorentheorie," which he wanted to finish first. In the meantime, Morgenstern wrote two chapters and an introduction. And in the diaries he stated that he wanted to integrate into the paper a remark on the measurement of utility (11 August 1941, OMPD).

In September 1941 von Neumann was able to concentrate on the manuscript for two weeks (8 September 1941, OMPD). At that point, the two agreed the paper should be expanded into a small book. The paper as written became the draft of chapter 1 of TGEB.[9] Von Neumann seemed

8. The expectation that the theory of games would have to overcome a lot of ignorance was a recurring theme during the whole work, i.e., "He (von Neumann) thinks that only a few can be convinced. Many will not *want* to hear, others will not understand. We are prepared for the most foolish objections and comments" (14 August 1941, OMPD).

9. "Johnny proposed to include it in the mathematical series, where he, Goedel, Weyl, etc.

to like the work. The more he got into the problems, the more the manuscripts were in his handwriting. This is the case for chapter 2, which was written in October and November of 1941. Morgenstern remarks, "We have been working hard. Johnny has the lion's share" (31 October 1941, OMPD). There is no indication that this changed throughout the rest of the writing of the book. Although the two often sat together to write, several times von Neumann wrote as much as seventeen pages overnight and calculated examples (11 April 1942, OMPD). Time and again Morgenstern was surprised by von Neumann's abilities and expressed his admiration, for example, when he axiomatized utility: "It developed gradually, faster and faster, and at the end, after two hours (!), it was almost completely finished. It was a great delight and touched me so much that afterwards I could not think of anything else" (14 April 1942, OMPD).

On 24 December 1941, Morgenstern reported the discovery of Ville's elementary proof of the minimax theorem, which he had found in a book by Borel: "Both are unknown to Johnny. Now he has discovered additional proofs that are becoming increasingly simple and are purely algebraic!! It necessitates some modification in the text, but we can print it" (24 December 1941, OMPD).

It seems that von Neumann was indispensable for the work. There was no progress without him. When he was absent or busy Morgenstern executed his work as a secretary typing the manuscript and putting in formulae (Morgenstern 1976a, 812 and, for example, 29 April 1942, OMPD). These were the times when he realized what difficult work they were producing together and how much trouble others would have understanding it.

Sometimes when von Neumann was writing, Morgenstern was thinking about insertions and remarks.[10] They reflect his main concerns and

---

had published. As a paper, it is much too long. I suggested we make a little book and have it published by the P.U. Press. That is what we will do now" (22 September 1941, OMPD). This statement reveals a conflict between the diaries and Morgenstern's memory. "As we continued to work, Johnny said: 'Why don't we go to Princeton University and ask them whether they would be interested in such a pamphlet?' " (Morgenstern 1976a, 809).

10. By this time, Morgenstern thinks that the book will have about 200–250 pages. Later, at the beginning of March, he already expects 400 pages (5 March 1942, OMPD). During the rapid development of the manuscript the title underwent several changes:

—"Theory of Rational Behavior and of Exchange" (22 September 1941)
—"Principles of Rational Economics" (31 October 1941)
—"Mathematical Theory of Economic Behavior" (24 December 1941)
—"Theory of Games—and Its Application to Economics and Sociology" (4 December 1942)

the extent to which these concerns (which were minimal) found their way into the text. I provide a short list of such topics:

—Morgenstern wrote a short paper about theoretical difficulties with demand curves (13 November 1941, 11 April 1942, OMPD). Initially, he wanted to integrate it as a note in the book, but they must have decided to treat the issue in a separate publication, which Morgenstern did years later (Morgenstern 1948).
—For the end of the section on poker, Morgenstern announced his intention to emphasize the possibilities of representing psychological processes in mathematical terms (21 January 1942, OMPD). However, there is not much left of this in the TGEB.
—On 29 March 1942, Morgenstern reported that he had outlined a good deal on general equilibrium, so as to show more comprehensively that, instead of a problem of maximization, there is a minimax problem.[11] This section in the TGEB is not very detailed.
—Just at the beginning of the actual collaboration, Morgenstern remarked: "But it will take a long time before the idle talk on dynamic theory in our times will cease" (12 July 1941, OMPD). This indicates that in the beginning Morgenstern focused on a solution for dynamic rather than static economics. Some months later it sounded a bit more vague: "We still have not thought about one of my points for chapter II: what is the difference between static and dynamic games" (31 October 1941). What remains in the TGEB of these concerns about time sounds like an excuse for the theory's being static (von Neumann and Morgenstern 1953, 44–45).
—Once, he acknowledged contentedly: "My proposal to understand the v(s) function as a vector was accepted and we inserted a paragraph in chapter VI" (25 May 1942, OMPD).
—On 17 October 1942, Morgenstern noted that he was thinking about the economic examples with which the book would end (OMPD).
—The measurement of utility was a particularly important theme for Morgenstern. He had to convince von Neumann of the importance of the subject and make him work on it (letter, 13 October 1942, OMP, Box 51, subject file "Neumann, John von, 1940–1945").

11. The need for the more comprehensive treatment of this problem arises because many economists, e.g. Myrdal, make the objection that under the condition of free competition it would really be a maximization problem because, by definition, all the data are given (17 June 1942, OMPD).

In September 1942, von Neumann moved to Washington to work for the U.S. Navy. This limited their opportunities to work together. Apart from some joint discussions on weekends, von Neumann apparently did most of the rest alone. This is clearly true for the part about non-zero-sum games. In a letter to von Neumann, Morgenstern wrote, "I am sorry that again I cannot be helpful but yet I am curious about how the manuscript is going to look" (7 November 1942, OMP, Box 51, subject file "Neumann, John von, 1940–1945"). At the end it appears that they did not have much time left for the economic examples. Both were very happy when they discovered that their results at least reasonably corroborated existing knowledge in addition to providing some new insights. On 28 December 1942, von Neumann's thirty-ninth birthday, they managed to finish the manuscript; this was just in time, for at the beginning of 1943 von Neumann left on a military mission to England.

It can be concluded that Morgenstern made only a minimal contribution to the theoretical and technical side of the innovation. There is no reason to mistrust his own judgment that he had a "minimal share." The mathematical formulation of the TGEB was clearly von Neumann's business. The theory of games represented his theoretical innovation. Only he was able to bring the theory into the form it took in the TGEB. One aspect of this asymmetry is that Morgenstern's diaries do not show even minor disagreement with von Neumann. One gets the impression that the static nature of the theory was the only point where Morgenstern's agenda was not satisfied. But Morgenstern did not oppose much; he accepted the limitations, although the importance of the element of time had been one of his major concerns during the 1930s. He was dazzled by von Neumann's mathematical abilities, but who would not have been?

Even if Morgenstern's immediate contribution to the technical side of the TGEB was small and he very often acted only as secretary, I think Niehans's (1990, 395) description of Morgenstern's role in the collaboration as an innovating entrepreneur is accurate. At von Neumann's side, Morgenstern, with his enthusiasm, propelled the innovator's project, and he provided in the first chapter the economic context for the theory of games.[12] There and in his diaries it becomes clear that he saw the theory of games as a challenge to neoclassical economics.[13] For fur-

12. It is, of course, another story that the theory of games did not find their first context in social science but in the context of military research. The story is told by Leonard (1991) and Mirowski (1991).

13. See Mirowski (1992).

ther understanding of this point, one would need a careful analysis of the intellectual development of Oskar Morgenstern, but such a study is beyond the scope of this article.[14]

Appendix
Collaboration: Chronological Survey

The survey is divided into the following three phases:

1. Prelude, 1928–April 1940
2. Intense discussions, April 1940–June/July 1941
3. Actual collaboration, June/July 1941–December 1942 (1944)

First Phase: Prelude

1928	Von Neumann's (VN) Paper "Zur Theorie der Gesellschaftsspiele," including proof of the minimax theorem, is published. He promises a supplement with numerical examples, e.g., on poker.
1935	Morgenstern (OM) presents his paper on perfect foresight in Menger's mathematical colloquium in Vienna. The mathematician Eduard Cech draws his attention to VN's 1928 paper on games (Morgenstern 1976a, 806).
1937	Report that VN studies games such as poker for "mere recreation" (*Science News*, April 3, p. 216).
1938	OM visits the U.S. as a visiting professor. He learns that he is dismissed as head of the trade cycle institute in Vienna after the Anschluss. He decides to go to Princeton as a lecturer.
Fall, 1938	VN and OM meet for the first time in Princeton, "soon after the University opened" in the fall of 1938 (Morgenstern 1976a, 807).
February 1939	Second meeting: tea with Niels Bohr.
October 1939	First intense discussion between VN and OM, no immediate consequences; OM is thinking about "maxims of behavior."

14. I think it is essential to see Morgenstern's interest in the theory of games in the context of his whole intellectual development and not only in the light of his Sherlock Holmes example. My dissertation is a step in this direction (Rellstab, forthcoming).

Until April 1940	VN and his wife invite OM to three dinners, no report on specific discussions (25 January 1940, 9 March 1940, 1 April 1940, OMPD).

Second Phase: Intense Discussions

April 1940	The two minds really meet; several intensive discussions, on behavior, games, mathematics, etc. OM gets the impression that VN is interested in his work. OM speculates in his diaries on a joint book with VN, who leaves for a trip to the West Coast.
Summer 1940	In Seattle, VN lectures on games and makes progress with four-and-more-person games.
August 1940	The two meet again after VN's return; the intellectually intense relationship continues.
September 1940	Familiar terms: OM refers to von Neumann as "Johnny" in his diaries.
October 1940	VN writes a first paper on games ("General Foundations"); contents: value function, solution, domination. VN lectures on games in Princeton.
January 1941	VN writes a second paper on games ("Decompostition Theory").
May 1941	VN advises OM to write a paper on "maxims of behavior"; OM begins immediately.

Third Phase: Actual Collaboration

June–July 1941	VN proposes to write a paper on games together; OM begins an introduction to VN's papers on games.
August 1941	The result of OM's work is an attempt to simplify VN's theory. It appears that this is the paper for economists OM refers to in 1976a. (I was not able to locate this draft, but there is absolutely no indication in the diaries that OM wrote the paper before the joint decision to collaborate.)
September 1941	Intense collaboration between VN and OM. Results: draft of the first chapter of the TGEB; *OM* (not VN; Morgenstern 1976a, 809) proposes they write a book together.
October–November 1941	Draft of chapter 2: VN does the lion's share, characteristic of the rest of the work.

December 1941	OM finds Ville's proof of the minimax theorem.
September 1942	VN moves to Washington; the collaboration becomes more difficult. There is not much time left for the economic examples. Only minor results.
28 December 1942	The manuscript is finished.
January 1943	VN leaves (for England), OM organizes the typing and the drawing of the graphs.
April 1943	OM delivers the manuscript to the Princeton University Press. John D. Rockefeller, 3d, financially supports the publication of the book.
1944	Publication of *The Theory of Games and Economic Behavior*.

References

Leonard, R. J. 1992. Creating a Context for Game Theory. *HOPE*, this issue.

Mirowski, P. 1991. When Games Grow Deadly Serious: The Military Influence on the Evolution of Game Theory. In *Economics and National Security*, edited by C. D. W. Goodwin. Durham, N.C.: Duke University Press.

———. 1992. What Were von Neumann and Morgenstern Trying to Accomplish? *HOPE*, this issue.

Morgenstern Papers, Oskar. Duke University Library. Durham, N.C. (OMP).

Morgenstern Papers: Diaries, Oskar. Duke University Library. Durham, N.C. (OMPD).

Morgenstern, Oskar. 1928. *Wirtschaftsprognose*. Vienna: Springer.

———. 1948. Demand Theory Reconsidered. *Quarterly Journal of Economics* 62 (February): 165–201.

———. 1959. *International Financial Transactions and Business Cycles*. Princeton: Princeton University Press.

———. 1976a. The Collaboration between Oskar Morgenstern and John von Neumann on the Theory of Games. *Journal of Economic Literature* 14.3:805–16.

———. 1976b. *Some Thoughts on Maxims of Behavior in a Dynamic Universe: Mélanges en l'Honneur d'Henri Guiton*. Paris: Dalloz.

Niehans, J. 1990. *A History of Economic Theory: Classic Contributions, 1972–1980*. Baltimore: Johns Hopkins University Press.

Rellstab, U. (forthcoming). Der Beitrag Oskar Morgensterns zur "Theory of Games and Economic Behavior" von Neumann und Morgenstern. Ph.D. diss. University of St. Gallen.

Schotter, A. 1976. *Selected Economic Writings of Oskar Morgenstern*. New York: New York University Press.

———. 1992. Oskar Morgenstern's Contribution to the Development of the Theory of Games. *HOPE*, this issue.

Science News Letter. 1937. Princeton Scientist Analyzes Gambling: You Can't Win. April 3.
Shubik, M. 1981. Oskar Morgenstern. In *Essays in Game Theory*, edited by R. J. Aumann. Mannheim: Bibliographisches Institut.
von Neumann Papers, John. Library of Congress. Washington, D.C. (VNP).
von Neumann, J., and O. Morgenstern. [1944] 1953. *The Theory of Games and Economic Behavior*, 3d edition. Princeton: Princeton University Press.

Oskar Morgenstern's Contribution to the Development of the Theory of Games

Andrew Schotter

The theory of games represents one of the few genuine social-scientific inventions of the twentieth century. Unlike many contributions to social science which rely on or are applications of some preexisting theory, game theory was developed by von Neumann and Morgenstern *ex nihilo*. Despite some similar work by Émile Borel, there are almost no precursors to the book *The Theory of Games and Economic Behavior* (hereafter called *The Theory*). To be assigned the task of measuring Oskar Morgenstern's contribution to this invention is like being a journalist asked to investigate James Crick's contribution to the discovery of the double helix after both Watson and Crick had died. This is a difficult task since intellectual imputation among scholars is very awkward, especially when an "imputee" is one of the twentieth century's greatest minds. Despite this difficulty, however, I will begin my essay by discussing the topic assigned to me in its narrow form, What was Oskar Morgenstern's contribution to the development of the theory of games? As a corollary I will discuss some subsidiary topics such as, How would game theory be different if someone else had collaborated with von Neumann? What future events in game theory and economic theory were anticipated by von Neumann and Morgenstern in 1944 and what developments would Morgenstern have been most excited about if he were still alive today?

Before I start, let me list my credentials for this assignment. First of all, *The Theory* was written in 1944, three years before I was born. Hence, it should be obvious that my insights come not at all from personal observation. I cannot claim to have been some invisible third person sitting in von Neumann's livingroom while he and Morgenstern

This paper was written for presentation at the Conference on the History of Game Theory held 5–6 October 1990 at Duke University. The author would like to thank William Baumol for his many comments on the paper as well as the technical support of the C. V. Starr Center for Applied Economics.

wrote their famous book. In fact, quite the opposite is true. I knew Oskar Morgenstern at New York University after his retirement from Princeton and up until his death. I was in some sense his last student. But his lectures in the 1970s clearly had the same zeal and enthusiasms as they must have had in the 1950s (in fact I am quite sure they were the same lectures since we only studied *The Theory*). Hence, in some indirect way I feel that I too was there in the beginning and was privileged to insights into his relationships with von Neumann and others.

Oskar Morgenstern's Contribution to the Development of Game Theory

Contributing to *The Theory*

As Morgenstern states in his essay on collaborating with von Neumann (Morgenstern 1976) one of the questions most asked of him in the years after von Neumann's death was what exactly was his contribution to the theory of games and to their collaboration. Clearly, when one collaborates with a recognized genius one is assumed to be a junior partner until proven otherwise. In his essay, Morgenstern spells out the answer in great detail; I will discuss it shortly. Whenever I have been asked the question of Morgenstern's contribution, my response has been to ask the questioner what he or she thinks he or she would produce if he or she had access to John von Neumann for one year. Upon reflection, the response is usually silence since the questioner soon realizes that having access to von Neumann is no guarantee that one would be able to interest him in any problem, let alone produce something as grand as *The Theory*. Quite simply, Oskar Morgenstern was a visionary constantly on the lookout for the new and the unusual. Unlike many visionaries who ultimately are looked upon as kooks, Morgenstern was capable of transforming his vision into tangible and pathbreaking works. His freedom from the strictures of Anglo-American neoclassical training (he was a product of the Austrian School) allowed him to see a new path for mathematical economics, one freed from the physicslike maximizing models that Samuelson and others were creating. It was this visionary quality that combined so well with von Neumann's mathematical abilities to produce a new tool and a new field of inquiry.

In concrete terms it is very simple to list Morgenstern's contribution to *The Theory*. To begin, as the book's title explains, there are two parts to the 1944 classic—game theory and economics. As must be obvious, the pure game theory contained in these pages was clearly a creation of von Neumann's, although I think that his choice of topics, especially on the cooperative game side, was heavily influenced by Morgenstern. In terms of economics, however, *The Theory* was a natural outgrowth of several earlier ideas of Morgenstern's and must be appreciated as a milestone in the evolution of Austrian Economics.

To begin, in his early work on economic forecasting, *Wirtschafts-prognose* (1928) and "Volkommene Voraussicht und Wirtschaftliches Gleichgewicht" (1935) (translated by Frank Knight as "Perfect Foresight and Economic Equilibrium" in Schotter 1976), Morgenstern clearly saw the problem of strategic interaction among economic agents as the central problem and the individual maximizing model of neoclassical economics as an inadequate representation of it. For him, economics consisted of both "dead" and "live" variables in which the dead variables' values were determined by nature while the live ones' were determined by their actions and the actions of others. In order for an agent to decide how to behave rationally in such circumstances, that agent must know how others are expected to behave, but these actions involve a similar expectation on the part of others. Hence, Morgenstern saw the perfect foresight assumption, so critical to the general equilibrium theory of the time, as a contradiction. If people had perfect foresight into others' actions, then unless all of these expected actions actually formed an equilibrium, someone would deviate and not behave as expected to. This idea was to resurface later under the banner of the Rational Expectations Equilibrium, and I will return to that point later on. The work of Morgenstern is probably the first statement of the rational expectations problem. Such a situation seemed circular to Morgenstern but was clearly the essence of "interesting" social science. This circularity was expressed most clearly by Morgenstern in his example of Sherlock Holmes and Professor Moriarty, presenting probably the first instance of a game with a mixed strategy equilibrium to appear in an economic article.

> Sherlock Holmes, pursued by his opponent, Moriarty, leaves for Dover. The train stops at a station on the way, and he alights there rather than travelling on to Dover. He has seen Moriarty at the railway station, recognizes that he is very clever, and expects that Moriarty will take a special faster train in order to catch him at Dover. Holmes' anticipation turns out to be correct. But what if Moriarty had been still more clever, had estimated Holmes' mental abilities better and had foreseen his actions accordingly? Then obviously he would have

> travelled to the intermediate station. Holmes, again, would have had to calculate that, and he himself would have decided to go on to Dover. Whereupon Moriarty would have "reacted" differently. Because of so much thinking they might not have been able to act at all or the intellectually weaker of the two would have surrendered to the other in the Victoria Station, since the whole flight would have become unnecessary. Examples of this kind can be drawn from everywhere. However, chess, strategy, etc. presuppose expert knowledge, which encumbers the example unnecessarily. (Morgenstern 1935, reprinted in Schotter 1976, 174)

For someone with these interests, the theory of games was a natural end product and, in fact, it was precisely this work that first stimulated the collaboration between von Neumann and Morgenstern.

Morgenstern was a visionary, and his vision can be seen most clearly in the introduction to *The Theory*. In this introduction three major things are accomplished. First the problem for economic science is shifted from a neoclassical world composed of myriad individual Robinson Crusoes existing in isolation and facing fixed parameters against which to maximize, to one of a society of many individuals, each of whose decisions matter. The problem is not how Robinson Crusoe acts when he is shipwrecked, but rather how he acts once Friday arrives. This change of metaphor was a totally new departure for economics, one not appreciated for many years. Second, the entire issue of cardinal utility is discussed, a problem whose importance to economics was clearly known to Morgenstern. While the proof of the existence of a cardinal utility was most certainly left to von Neumann, who thought the whole thing quite obvious, the axiomatization was obviously influenced by Morgenstern. An attempt is made to keep the axioms as close as possible to those needed to prove the existence of an ordinal utility function under certainty—a concern that Morgenstern must have felt more strongly than von Neumann. Finally, the entire process of modeling exchange as an *n*-person cooperative game and searching for a "solution" is described. Here one is struck by how Austrian was the economics used in *The Theory*. The presentation is clearly motivated by Böhm-Bawerk's example of a horse market—a market in which the "marginal pairs" determine equilibrium price. While the neoclassical theory of price formation was calculus-based and relied on first-order conditions to define equilibrium, game theory, especially cooperative game theory, relied more on solving systems of inequalities (such as the inequalities determining the core). While the neoclassical solution would often be included within the set of cooperative solutions (and would often prove to be the limiting solution as the number of agents in the market approached infinity), the theory of games offered new and appealing *other* solutions to the problem of exchange. This non-neoclassical emphasis, I feel, is another of Morgenstern's contributions to the development of game theory.

Finally, it is clear that Morgenstern saw game theory, or at least the von Neumann–Morgenstern (stable set) solution, as a formalization of a revived "neo-institutional" economics. Such an institutional emphasis was, of course, seen in earlier work by Menger (1883) where he investigates the organic or unplanned creation of social institutions. For Morgenstern the institutional question arises because he saw game theory as a tool to allow social scientists to define the set of possible, mutually exclusive, institutional arrangements that could emerge from a given social situation. If you substitute the words "institutional arrangements" for "orders of society" in the quotation below you will see the point being made.

> The question whether several stable "orders of society" based on the same physical background are possible or not, is highly controversial. There is little hope that it will be settled by the usual methods because of the enormous complexity of this problem among other reasons. But we shall give specific examples of games with three or four persons, where one game possesses several solutions in the sense of 4.5.3 [the stable-set solution]. And some of these examples will be seen to be the models for certain simple economic problems. (von Neumann and Morgenstern 1947, 43)

To give an example of exactly how Morgenstern saw the institutional aspect of game theory and how he envisioned its being different from the standard neoclassical analysis, consider the following example taken from an early draft of a paper by von Neumann and Morgenstern (the only paper I know of on which they collaborated after the *Theory of Games* was written) which was left unfinished, but completed almost thirty years later by Morgenstern and Schwödiauer (1976).[1]

Consider a simple partial-equilibrium analysis of a market in which there is one seller who has two units of a good to sell and three buyers, all

1. The analysis presented here is also described in Braunstein and Schotter 1978.

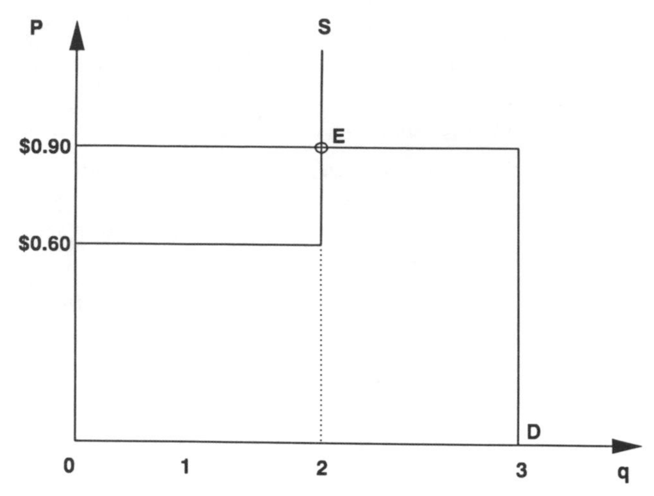

Figure 1

of whom want to buy one unit. Further assume the seller will sell any unit for a price of $0.60 or more, while each buyer is willing to buy any unit for a price up to, but not exceeding, $0.90. With this information, the conventional analysis of price formation would proceed by constructing the appropriate supply and demand curves and solving for that price at which supply equals demand (see figure 1).

In the neoclassical analysis the question is asked, What price or value relationships are consistent with the hypothesized institutional structure of perfect markets? The emphasis is on values given institutions. Consequently, since the buyers' side of the market is the long side, the price is supposedly bid up to the buyers' reservation price, yielding an imputation in which the seller receives $0.30 profit on each unit sold, while the buyers' consumer surplus is reduced to zero or an imputation of $x = (60, 0, 0, 0)$. The market price, $0.90, is associated with this imputation, and if the analysis is accepted as valid, an equilibrium institution-price relationship is established.

But is this really the relevant question? In other words, is the relevant question what the equilibrium price is, consistent with the assumption that perfect markets exist, or should we, as social scientists and economists, be asking which equilibrium institution-price pair will emerge as the stable pair, or set of pairs, in this situation of strategic and social interdependence. If it turns out that the only stable institution-price pair is the competitive market-competitive price pair (which, by the way, is a feature of Lucas's [1968] counter-example demonstrating the nonexistence of a stable-set solution to a ten-person game which Shapley and Shubik [1969] in turn demonstrate to be a market game) then we can say that the neoclassical analysis is not institutionally myopic in its assumptions. But if institutions other than competitive markets might emerge and define different equilibrium value relationships which may involve a set of personalized (hence not competitive) prices, then the neoclassical approach is myopic in not being able to define these equilibrium institutions and their associated pairs as logical possibilities.

To make this point more precisely, Morgenstern and Schwödiauer look at the cooperative game-theoretical analysis of the exchange situation described above. To do this they must define the characteristic function associated with this situation by defining the value of any coalition as the maximum sum of the consumers' and producers' surplus that can be achieved by any coalition of traders. Letting the seller have the index 1 and the buyers the indices 2, 3, and 4, and assuming transferable utilities and side payments, the characteristic function appears as

$$V(1) = V(2) = V(3) = V(4) = 0$$

$$V(12) = V(13) = V(14) = 30, V(23) = V(24) = V(34) = 0$$

$$V(123) = V(124) = V(134) = 60, V(234) = 0$$

$$V(1234) = 60$$

Now if buyers restrict their behavior to coalition-forming behavior or "blocking"-recontracting behavior, then the equilibrium imputation associated with this behavior is the unique core imputation $x = (60, 0, 0, 0)$ in which the seller sells both units for a price of \$0.90, thereby extracting all of the consumer surplus from the buyers. This outcome is then identical with the outcome attained by the neoclassical market. However, if one wanted to, in this case, one could say that this blocking or recontracting behavior defines a competitive market–competitive price pair as the equilibrium institution-price pair for this situation of exchange. But traders need not restrict themselves to this type of behavior. There may be other standards of behavior that lead to other stable institution-price or imputation pairs as defined by the von Neumann–Morgenstern solution, and these equilibrium institutions should not be ruled out a priori by the assumption that only markets of the competitive type exist in the economy under investigation. For instance, the buyers, realizing that blocking or recontracting will inevitably lead to the core, may form a cartel and refuse to bargain with the sellers except as a unit. If they do this, then there are several equilibrium-stable institution-price pairs that might emerge, as Morgenstern and Schwödiauer (1976) demonstrate. For instance, the following four sets of imputations collectively define the total symmetric sets of equilibrium institutional relationships. (Here x_2 is the buyer with the biggest imputation, x_3 the second biggest, and x_4 the smallest.)

The core-market: $X^1 = \{x \,|\, x = (60, 0, 0, 0)\}$. (1)
This is the unique core imputation yielding a price per unit of \$0.90. It is identical to the neoclassical market solution and hence we call it a market solution.

The two-trader symmetric cartel: $X^2 = \{x \,|\, x_1 = (60 - x_2 - x_3 - x_4, 15 \geq x_2 = x_3 \geq 0, x_4 = 0)\}$. (2)
Here two buyers form a cartel and exclude the third buyer. The set of imputations defined is one in which the price is reduced below its competitive market price of \$0.90 to any price between \$0.90 and \$0.75. All traders buy at the same price.

The three-trader symmetric cartel: $X^3 = \{(x \,|\, x_1 = 60 - x_2 - x_3 - x_4, 20 \geq x_2 = x_3 = x_4 \geq 16^{2/3})\}$. (3)
Here all three buyers form a cartel and bargain with the seller as a unit. Because of their collusion they are able to force the price into the interval between \$0.60 and \$0.65.

The three-trader asymmetric cartel with discriminated buyer: $X^4 = \{x \,|\, x_1 = \{(60 - x_2 - x_3 - x_4, 18^{1/3} \geq x_2 = x_3 = c \geq 15, 37^{1/3} - C \geq x_4 \geq 0, x_4 \in M(C))\}$. (4)
Here the three buyers form a cartel, bargain with the seller as a unit, but do not split the gains from collusion equally. Rather, one buyer is discriminated against, who we are assuming in this case is buyer 4, and is merely given a side payment for his cooperation.

Notice that all of these different standards of behavior or institutional arrangements are logical possibilities for what may emerge from the situation of primitive exchange described before, and that neoclassical theory misses the opportunity to predict the emergence of any institution other than the competitive market (in this case X^1), which it assumes to exist at the outset of the analysis. Hence, from an institutional point of view, the analysis must be considered myopic.

This emphasis on institutional analysis is distinctly Mengerian and most certainly a contribution by Morgenstern. It is probably no surprise that the work of Martin Shubik (Morgenstern's most famous student) has consistently been in what he calls "mathematical institutional economics." Nor is it surprising that he, in collaboration with his coauthors Lloyd Shapley and Pradeep Dubey, should have pursued models with very great institutional detail since that pursuit was clearly implied as the mission of game theory in *The Theory*. This connection of institutions and game theory has also been highlighted by Sugden (1986) and Schotter (1981) as well as by the more mathematical versions of Williamson's "new institutional economics."

In summation, *The Theory* was a natural outgrowth of the work of its two coauthors, one who pioneered the study of games of strategy and the other who cared greatly about the problem of strategic interaction in social and economic affairs.

Contributions after *The Theory*

What influence did Oskar Morgenstern have on the development of game theory after *The Theory* was written? By influence I mean only that influence created by his published work. As anyone familiar with Morgenstern will know, his influence in the field at the time came from many sources, not the least of which was his role as a mentor and catalyst for others. His Econometric Research Program at Princeton in the 1950s was a great meeting place where game theorists congregated and where game theory flourished. When a new field is created out of thin air, one of the ingredients necessary for the field to grow is a strong leader who will encourage others and arrange for opportunities for the best people to find positions and obtain research funding. Morgenstern had a long history as such an entrepreneur extending back to his days in Vienna where he ran the Institute of Business Cycle Research and, among other things, arranged for Abraham Wald to come to America. His leadership at Princeton is well known to those who passed through but will not be the focus of my survey here. I will concentrate on his published work only.

In this connection it is unfortunate, but in some sense not surprising, that Morgenstern did very little in the way of game-theoretical application in the years after the publication of *The Theory*. In fact, aside from some surveys and his work on the von Neumann model of an expanding economy with Thompson, Kemeney, and others, I cannot think of another article than the one mentioned above written with Schwödiauer, that can be considered either an application or an extension of the theory. Rather, Morgenstern went on to write in a wide variety of fields including the predictability of stock market prices, the von Neumann model of an expanding economy, the use and abuse of economic statistics, matters of national security, and many others. Hence Morgenstern's written influence on the development of game theory stopped with the publication of *The Theory*. Still, the agenda for the application of game theory to economics in the fifties, sixties, and seventies was skewed substantially toward the application of cooperative games. I feel this agenda was heavily influenced, at least indirectly, by Morgenstern through the work of Martin Shubik and Lloyd Shapley, the first and most influential authors to pick up the call to arms offered in *The Theory* in a manner consistent with its intent. While the noncooperative theory has become the main tool used today in economics, such an emphasis might have occurred earlier had not attention been focused on the core solution concept in the period up until the late 1970s. Further, although the book is entitled *The Theory of Games and Economic Behavior*, it was clearly conceived as a general theory for all social science. As Professor Riker discusses elsewhere in this volume, cooperative game theory was the major game-theoretic tool used by political scientists for at least forty years after *The Theory* was written. In fact, the analysis of simple games in *The Theory* already contained the central element for the study of voting games developed by Riker, Ordeshook, McKelvey, and others in the 1970s. It always was on Morgenstern's agenda to give social science a new, and more flexible, mathematics with which to study social phenomena, and the cooperative theory clearly seemed well suited for that purpose. In fact Morgenstern notes that von Neumann was always amazed at the primitive state of mathematics in economics; he commented in the 1940s that if all economics texts were buried and dug up one hundred years later, people would think that the economics they were reading was written in the time of Newton. This emphasis on cooperative game theory and the belief that the proceeds of bargaining, threatening, and bluffing are of central empirical importance to social and economic life stem directly from Oskar Morgenstern, and I think strongly bear his imprint.

Later Events Anticipated by *The Theory*

In this section I will attempt to outline the various later developments in the theory of games and its economic application that were anticipated by Morgenstern and von Neumann in their original work. Let me say, however, that my definition of the word "anticipated" here is rather broad since I include all work that is consistent, at least in spirit, with *The Theory*. Finally, in addition to outlining what was anticipated, I will also indicate what I feel were developments that could not be considered natural outgrowths of the work contained in *The Theory*.

Anticipated Materials

Clearly the work in the 1960s and 1970s on the core of an exchange economy is the most natural outgrowth of the analysis investigated in *The Theory*. Von Neumann and Morgenstern had already defined the characteristic function for analysis of markets, although they had clearly pinned their hopes on the stable-set solution. The idea of the core, a subset of every stable-set solution, was not foreign to them. It was, of course, left to others such as Martin Shubik, Herbert Scarf, Gerard Debreu, Lloyd Shapley, and Robert Aumann to demonstrate the relationship between the core and the conventional Edgeworth contract-curve analysis. Despite the elegance of this literature, the limit theorems on the core might still be considered a disappointment to Morgenstern. While he clearly wanted a theory that could include the competitive analysis as a special case, I think that he had hoped that something more than the competitive equilibrium in the limit would exist. The idea that as societies grow the only stable "orders of society" or institutional arrangements are competitive markets is rather inconsistent with the spirit of the stable-set analysis.

We have evidence that Morgenstern held such a view. In a later paper (Morgenstern 1949, published in Schotter 1976) he states,

> The economist's prime interest in the success of the extension of the number of players will naturally be whether the present basic belief that large numbers of buyers and sellers secure a free competition where each individual faces a pure maximization problem, finds any strengthening or not. As far as I, personally, can see there will be nothing of this kind, although some sort of asymptotic behavior of the solution may emerge. (298)

The Nash-Shapley-Harsanyi cooperative analysis of bargaining is another topic that naturally follows, at least in spirit, from *The Theory*. Bargaining is the prime mover of prices and allocations in the new game theory (in both the small- and large-numbers cases) and the cooperative theory of bargaining is consistent with this world view. In addition, the fact that early bargaining theories were axiomatized was, methodologically, a natural extension of the approach von Neumann and Morgenstern used in their utility analysis—an analysis that presented the first set of axioms in economics.

The axiomatization of fairness and its equivalence with the Shapley value was welcomed by Morgenstern. If bargaining was to be the prime mover of social interaction in the new theory of games, then it would have been natural for a corresponding moral theory to be developed that was based on the rationality of the players and on nothing else. As we will see later, to the extent that these notions serve a selection criterion of unique equilibria, I will show that such a consequence would not have been welcomed by Morgenstern.

Unanticipated Events

Little of what eventually happened to the field in the 1980s, especially the recent work in noncooperative game theory, was clearly envisioned in 1944, since *The Theory* did not pursue this analysis past its zero-sum minimax formulation. In fact, it is quite remarkable that the Nash equilibrium concept was not defined nor its existence proven in *The Theory* since at least Morgenstern was well aware of Cournot's work. I suspect the main reason for this omission is that both von Neumann and Morgenstern were looking for a way to break the circularity of the "I think he thinks that I think" logic of strategically interdependent situations. They wanted to provide a way for players to behave that was independent of their expectation of what their opponent intended to do. This is, in fact, exactly what the minimax strategy does since when one chooses a minimax strategy, one is *guaranteed* a certain return, no matter what one's opponent does. The Nash equilibrium does not provide this independence, since under such an equilibrium both parties must have correct conjectures about the actions of the other. Hence, it was not the type of solution they were looking for. Ultimately, however, they would have been driven to it since in most variable-sum games minimax strategies for players do not form an equilibrium. Von Neumann and Morgenstern also used the minimax formulation to define the characteristic function for a game, and did so for the reasons stated above. Their formulation ran into some conceptual problems that were later dealt with by the β and α definitions of the characteristic function and by Rosenthal's work on the effectiveness function forms for games.

Much of what we find interesting today in game theory and its application was not anticipated by (i.e., in the spirit of) *The Theory*. This is not to say that von Neumann and Morgenstern would not have found them interesting or important; it simply means that they were not directly in line with their thinking *at that time*. For example, the recent interest in refinements of Nash equilibrium, while probably exciting to Morgenstern personally, was not a natural outgrowth of *The Theory* given its lack of focus on the Nash concept. As noted before, the entire emphasis in *The Theory* was to find a decisive way to act that avoided any thought of what action one's opponent might take. Hence, elaborate analyses of beliefs off the equilibrium path were not consistent with this view.

Further, the extent to which refinements focus on choosing a unique outcome for each game is the extent to which that program is inconsistent with the world view expressed in *The Theory*. For Morgenstern, indeterminacy was not something to run from but rather to embrace.[2] The world is uncertain and social situations interesting only because they contain indeterminacies that many physical situations do not. To explain this belief, Morgenstern and Thompson (1976) present the following example.

> The great physicist W. Nerst, who established fundamental laws of thermodynamics, showed a long time ago the difference between physical mechanisms and social processes by the following devise: Assume a body to be in the position A upon which two forces 1 and 2 work as shown in the figure below [see figure 2]. Then it is well known that the body will move along the dotted resultant toward A′. But Nernst said, suppose the "body" is a dog and in (1) and (2) are two sausages, each exerting an influence on A. Clearly, the dog will not walk to Point A′ between (1) and (2). (Morgenstern and Thompson 1976, 9)

2. This view is not shared by John Harsanyi, who is a strong advocate of the proposition that social-scientific theories should have determinate predictions. Harsanyi (1976) states that the main reason why game theory had not, as of that date, found extensive application is that von Neumann and Morgenstern's approach "in general does not yield determinate solutions for two-person nonzero-sum games and for n-person games."

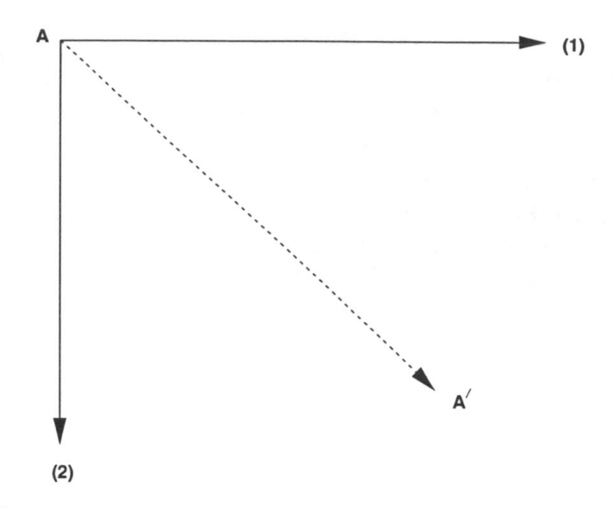

Figure 2

This is not to say that considerations of rationality should not be used to *narrow down* the set of possible equilibria that may emerge, but narrowing down is not the same as choosing one uniquely.

From an institutional view, the entire literature on mechanism design, while institutional in focus, concentrates more on the planned aspects of economic and social institutions and less on Menger's (and Morgenstern's) emphasis on unplanned orders of societies. I do not personally think that Morgenstern saw game theory as a tool for social planning and institutional engineering, although clearly this literature is some of the most exciting to be written in the past twenty years. The literature is too "un-Austrian" for Morgenstern.

Given his appreciation for the indeterminacy of social situations, I feel that Morgenstern also would have appreciated much of the work on repeated games (despite the fact that repetitions do nothing to change the analysis of zero-sum games). It is ironic, however, that he would have appreciated just that aspect of this literature that frustrates those who work in it—the large indeterminacy of the folk-theorem. For Morgenstern, it would not be surprising (or undesirable) that repetition of a super-game would lead to cooperation possibilities that were not available in one-shot play. For example, the emergence of cooperation in prisoners' dilemma games is clearly something that could be expected, particularly because one can view the play of repeated noncooperative games as a version of play of a cooperative one-shot bargaining game.

Finally, there is the modern literature on industrial organization, entry prevention, reputation building, signaling, and sequential equilibriums. One remarkable aspect of *The Theory* is its lack of attention to the noncooperative oligopoly problem despite, as mentioned earlier, Cournot's earlier work. In this sense one must place this literature in the unanticipated category (although the authors of *The Theory* would not have been surprised that their work was applied to the oligopoly problem. In fact this problem is mentioned on the very first page of the book, yet never presented anywhere as an example of the theory's potential applicability). Furthermore, many of these models are incomplete-information models in which the actions of incumbent firms are viewed as signals of their hidden costs. To the extent that bluffing possibilities exist in these models we see analyses that, in spirit, are very much like the bluffing analysis described by von Neumann and Morgenstern in the discussion of poker.

Most Appreciated Events

Since Morgenstern's study of strategic interaction was stimulated by his interest in the problem of perfect foresight and prediction, the recent work on rational expectations equilibria was a welcome event. This concept, in some sense, solves the problem that motivated Morgenstern's interest. Morgenstern's prediction problem is solved by Muth, Lucas, and others by assuming that all agents in the economy make their predictions using the same model of their situation. Hence, if all agents in an economy had the same "objectively correct model" and formed their forecasts of the future on this basis, the Rational Expectations Equilibrium forms a Nash Equilibrium in beliefs and actions. *At the equilibrium* the circularity of beliefs is settled and Morgenstern's original problem disappears. Still, the Austrian in Morgenstern would not have tolerated such a simplistic view of the world. For example, in the small-numbers case a Rational Expectations Equilibrium might be vulnerable to the problem of "theory absorption" discussed by Morgenstern and Schwödiauer (1976). Theory absorption asks whether a solution concept (or equilibrium notion) like the Rational Expectations Equilibrium or the core will be adhered to if people understand the theory behind it. For example, in a rational expectations model assume that I know that all agents are forecasting the price in the market using the same theory, and that it is common knowledge that all people are doing so and using these forecasts to guide their behavior. Further, assume that one of the parameters in the model is unknown and must be estimated using market data. If I am "large" enough I might try to manipulate the beliefs of others by deviating and acting as if I was forecasting prices using some other theory. This might then corrupt the data that others are observing and lead them to change their behavior to my benefit. Hence, knowledge of the rational expectations theory may, once absorbed, lead to a breakdown in the predicted equilibrium. When the number of agents in the economy is "large" such problems do not arise since no agent can corrupt the aggregate data by failing to adhere to the commonly used model. In fact, Morgenstern's original prediction problem must have been a small-numbers problem since even that problem would fail to exist with agents of measure zero. This same phenomenon of theory absorption was demonstrated to be serious for the core solution concept in an experiment done by Braunstein and Schotter (1978).

Beyond this, Morgenstern would probably have trouble agreeing with the assumption that all agents adhere to the same theory of the economy, a theory describing the objectively correct model of the economy. In fact, Morgenstern (1935) had already anticipated the rational expectations common-model solution and rejected it since economics even now is not yet a science for which we have a commonly accepted correct theory. He states, "Herein lies a contradiction which, in my opinion, can not be avoided for the aforementioned case, wherein full knowledge of the science, still not existing, has to be attributed to individuals because of full foresight" (Morgenstern 1935, in Schotter 1976, 176).

My feeling is that Morgenstern would have been more inclined to think of the agents in the world as adhering simultaneously to many theories and to think, in truly Austrian fashion, that many subjectively correct models of the real world exist, reality being determined, in part, by the different subjective models that people use. A sunspot model might even be closer to the type of analysis he might have envisioned. Hence, while in some sense the theory of rational expectation equilibrium would have been a very welcome event for Morgenstern since it dealt with precisely the problem that first aroused his interest in game theory, its treatment in the profession might ultimately have left him dissatisfied.

Although nothing in *The Theory* indicates this, the modern growth of experimental economics and experimental game theory was an "anticipated event." In a later paper (1954) Morgenstern commented on the already-published experiments of Chamberlain (1948) and clearly spelled out the possibility of reforming economic experiments. Since this topic is covered in depth by Vernon Smith, I will not pursue it further except to say that the model of physics experiments was one firmly planted in Morgenstern's mind, and as he explains in Morgenstern 1954 it is not a great leap of imagination to expect experiments to be done in economic and game theory.

Finally, since von Neumann was a pioneer in the development of computers and automaton theory and had influenced Morgenstern in this direction, one would think that the use of automatons as players in games would have been a welcome event. Clearly, on a personal level it would have been. The extent to which automatons represent less-than-fully-rational players with limited memories, however, is the extent to which the original spirit of *The Theory* may be violated. Still, it is hard for me to think that Morgenstern would not have welcomed the introduction of bounded rationality into games, perhaps along the lines of Radner 1987 or along the lines of Rubinstein 1986 and Neyman 1985. *The Theory* was simply too much of a first step to have taken things this far. In fact, the calculation requirements for the players in the types of games studied in *The Theory* is quite limited. Since they rarely get beyond four-person games, it is likely that when the combinatorial and calculating magnitudes of the problems got big, they too would have resorted to some type of bounded rationality premise to make sense of the situation.

Conclusion

Without Oskar Morgenstern we would not have the theory of games as we know it today. That does not mean that someone else would not have taken up work on von Neumann's problem, but it does mean that game theory would probably not have been introduced into the social sciences until many years later. Despite von Neumann's technical powers, the course of economics was changed by Morgenstern's focusing attention on a mode of analysis that has only recently come to be the dominant mode for all economics. The theory of games needed a non-neoclassical leader since it represented a fundamental break in economic thinking that would have been ruined if placed in the hands of a more conventional mind. The importation of Austrian imputation theory set the agenda for research in the 1960s and 1970s, and, whatever one thinks of the literature on the core, it is directly attributable to the change of emphasis initiated by Morgenstern in *The Theory*. I can think of no other economist at the time who could have walked into a room with John von Neumann and walked out later with a 600-page book on the theory of games complete with economic examples. That fact speaks for itself.

References

Braunstein, Yale, and Andrew Schotter. 1978. An Experimental Study of the Problem of "Theory Absorption" in N-Person Bargaining Situations or Games. In *Coalition-Forming Behavior*, edited by Heinz Sauermann. Tübingen: J. C. B. Mohr.

Chamberlin, Edward. 1948. An Experimental Imperfect Market. *Journal of Political Economy* 56 (April): 95–108.

Harsanyi, John. 1976. *Rational Behavior and Bargaining Equilibrium in Games and Social Situations*. Cambridge: Cambridge University Press.

Lucas, William. 1968. A Game with No Solution. *Bulletin of the American Mathematical Society* 74 (March): 237–39.

Menger, K. [1883] 1963. *Untersuchungen über die Methode der Sozialwissenschaften und der politischen Ökonomie insbesondere*, translated by Francis J. Nock as *Problems in Economics and Sociology*. Urbana: University of Illinois Press.

Morgenstern, Oskar. 1935. Perfect Foresight and Economic Equilibrium. *Zeitschrift für Nationalökonomie* 6.3 (August): 337–57.

———. 1949. Economics and the Theory of Games. *Kyklos* 3.4:294–308.

———. [1954] 1976. Experiment and Large-Scale Computation in Economics. In *Economic Activity Analysis*. New York: John Wiley & Sons.

Morgenstern, Oskar, and Gerhard Schwödiauer. 1976. Competition and Collusion in Bilateral Markets. *Zeitschrift für Nationalökonomie* 36.3-4:217–45.

Morgenstern, Oskar, and Gerald Thompson. 1976. *Mathematical Theories of Expanding and Contracting Economies*. Lexington, Mass.: D. C. Heath & Co.

Neyman, Abraham. 1985. Bounded Complexity Justifies Cooperation in the Finitely Repeated Prisoners' Dilemma. *Economic Letters* 19:227–29.

Radner, Roy. 1987. Can Bounded Rationality Resolve the Prisoners' Dilemma? In *Contributions to Mathematical Economics*, edited by A. Mas-Colell and W. Hildenbrand. Amsterdam: North Holland.

Rubinstein, Ariel. 1986. Finite Automata Play the Repeated Prisoners' Dilemma. *Journal of Economic Theory* 39.1 (June): 83–96.

Schotter, Andrew. 1976. *Selected Economic Writings of Oskar Morgenstern*. New York: New York University Press.

———. 1981. *The Economic Theory of Social Institutions*. Cambridge: Cambridge University Press.

Shapley, Lloyd S., and Martin Shubik. 1969. On Market Games. *Journal of Economic Theory* 1.1 (June): 267–79.

Sugden, Robert. 1986. *Economics of Rights, Cooperation, and Welfare*. Oxford: Basil Blackwell.

von Neumann, John, and Oskar Morgenstern. [1944] 1947. *Theory of Games and Economic Behavior*, 2d ed. Princeton: Princeton University Press.

What Were von Neumann and Morgenstern Trying to Accomplish?

Philip Mirowski

> I can play with the chessmen according to certain rules. But I can also invent a game in which I play with the rules themselves. The pieces in my game are now the rules of chess, and the rules of the game are, say, the laws of logic. In that case I have yet another game and not a metagame.—Ludwig Wittgenstein, *Philosophical Remarks*

> Turing's 'machines.' These machines are in fact human beings who calculate.—Wittgenstein quoted in Shankar 1987b

For most of the good half-century after it was published in 1944, *The Theory of Games and Economic Behavior* by John von Neumann and Oskar Morgenstern was more often cited in reverence than actually read. Its place in intellectual history was not considered a burning issue given that the status of game theory in departments of economics and mathematics was unclear. This has all changed, however, with game theory becoming established as a mathematical technique applicable to evolutionary biology, political science, and even various proof techniques within mathematics, as well as becoming the *nouvelle vague* in neoclassical micro theory (Kreps 1990). As game theory gains in stature, a nascent curiosity has surfaced concerning its origins and motivations, particularly as it now seems appropriate for various advocates to step back and ask such questions as "What is game theory trying to accomplish?" (Aumann 1985). It may serve as a propaedeutic for such methodological inquiries to first ask what it was von Neumann and Morgenstern

The author would like to thank Neil de Marchi, Roy Weintraub, and Craufurd Goodwin for all their help in facilitating the consultation of the Oskar Morgenstern archives (OMP) at Duke; Urs Rellstab and Robert Leonard for discussions concerning the papers of von Neumann and Morgenstern, and Pam Cook for all her patient translations from the German. The companion piece to this paper is my article (1991) describing the early response to the program of von Neumann and Morgenstern.

themselves were trying to accomplish, and to attempt to situate them within the larger intellectual currents of their era. This will open up the discussion to broader themes and variations not generally associated with game theory, such as the situation of artificial intelligence and formal logic in the first half of the twentieth century, as well as the possibility that game theory was not originally intended as a complement to the neoclassical research program. In fact, it may also raise the prospect that this joint-authored text might not always be singing in a contrapuntal voice, but instead modulate between different keys with little in the way of standard overture or cadenza.

The World according to the Orthodox Game Theorist

While it is premature to claim there is a standard or definitive history of game theory, there is a certain orthodox narrative used to set the stage for modern work.[1] It goes something like this: although various aspects of strategic economic behavior were formulated in a preliminary manner by such authors as Antoine-Augustin Cournot, Francis Ysidro Edgeworth, and Frederick Zeuthen, the first serious specification of the mathematical definition of pure and mixed strategies as well as the minimax solution was proposed by the mathematician Émile Borel in the years 1921–27 (Fréchet 1953). John von Neumann in a lone 1928 paper proved the existence of the minimax solution for all zero-sum games by means of the introduction of the innovation of the "characteristic function," which reduced the *n*-person game to its two-person counterpart through assignment of a value to each and every coalition and its complement.

There were no further developments in the theory until the appearance of *The Theory of Games and Economic Behavior* (TGEB), coauthored with Oskar Morgenstern in 1944. That book both axiomatized all the relevant ideas of game theory, from the formal structure of games of strategy to "von Neumann–Morgenstern utility," as well as providing an exhaustive discussion of zero-sum games. (The fact that TGEB claims to deal with non-zero-sum games is often ignored in this narrative.) It also spelled out the program of the application of game-theoretic ideas to

1. This narrative is cribbed from the following sources: Bianchi and Moulin 1991, Aumann 1985, Dimand and Dimand 1992, as well as various remarks made *en passant* by game theorists.

economic theory. However, some (unspecified) hostility to game theory retarded its advance within economics in the early years, causing most of the subsequent development to be located within military/strategic settings, or else under the rubric of "operations research." The former saw the development of noncooperative games, whereas the latter stressed the equivalence between linear programming and the computation of optimal mixed strategies. Von Neumann himself was too busy with the atomic bomb and early computers to lend the program much attention up until his death in 1957; and the subsequent role of Morgenstern is largely *terra incognita*.

In any event, the next landmark in this history is John Nash's 1951 paper on a solution concept for a class of non-zero-sum games dubbed "non-cooperative." This "Nash equilibrium" grew in significance in economics beginning roughly in 1975 to the point that today it is the basis for almost all work on the refinements of equilibrium concepts, such as the Harsanyi-Selten tracing procedure. On the other hand, the "core" of a cooperative game was defined in 1959 and shown by Debreu and Scarf in 1963 to converge to a Walrasian general equilibrium, thus demonstrating that neoclassical theory and the game-theoretic program were perfectly compatible.

The next "breakthrough" in the development of game theory according to this narrative was the 1967 paper by John Harsanyi which argued that games of incomplete information were in principle no different from games of complete information, primarily by stressing a Bayesian conception of "learning" on the part of the players. This in turn led to all manner of assertions about "dynamic" games, games with random perturbations, games of reputation, and so forth. Consequently, new equilibrium concepts began to proliferate, to such an extent that the present rate of invention is at least one a year, and accelerating. Some equilibrium concepts lost favor along the way, such as the kernel, the bargaining set, and von Neumann's own "stable set"; but by the 1990s, it has become apparent that the birth rate now far outpaces the death rate, with the population of equilibrium concepts nowhere near their Malthusian saturation point.

The finale of this orthodox narrative is a bit more elusive. The troublesome implications of this population explosion are generally avoided, with the game theorist instead stressing the obvious successes of game theory in coming to dominate whole sections of the fields of industrial organization, macroeconomics, and even the low-rent district of comparative economic systems. If an explicit narrative of progress is called for, one is sometimes located in what has been called the "Nash Program" (Binmore and Dasgupta 1987), which asserts that all cooperative games can be reduced to their noncooperative counterparts by explicitly modeling the pregame negotiations; and then moving on to assert that every new innovation in equilibrium concepts is in some shape or form a "refinement" of the Nash equilibrium point. Yet even the proponents of this position hesitate to endorse the project of Harsanyi and Selten (1988) to have found the one true unique equilibrium (Binmore 1989).

It is my opinion that this narrative is riddled with gaps, silences, longueurs, embarrassments, and inconsistencies. It takes for granted that the game-theoretic program of research is more or less consistent with orthodox neoclassical theory and that its progressive character is not contentious even among game theorists, and it conveniently overlooks the fact that some prominent orthodox theorists have expressed persistent hostility to the program over their entire careers.[2] It bundles together many writers who make uncomfortable bedfellows in the context of the history of economic thought and presumes that mathematical innovation is identical to economic inventiveness. Worst of all, it is blind to the fact that the most mathematically sophisticated proponent of the program until very recent times, John von Neumann himself, did not regard economics as the focal point of its development. We can begin to rectify this situation by training our sights upon the progenitors, first individually, and then as a coalition.

Reducing Epistemological Enigmas to Mathematical Formalisms: John von Neumann (1903–1957)[3]

More than under any other single mantle, most of the diverse projects of John von Neumann can be understood as reactions to the Hilbert pro-

2. "John von Neumann and Oskar Morgenstern had complementary properties that led to a fruitful co-operative product. . . . But, it can be said in a low-keyed way, there was something in their respective scientific personalities that led to a resonance of minor faults—some tendency towards nihilism combined with Napoleonic claims. Schumpeter was pleasantly excited by John von Neumann's claim that economics would need new mathematics different from the mathematics developed for physics after Newton's time. . . . However, except for the philosophical complications introduced by games involving more than one person, I do not honestly perceive any basic newness in this so-called non-physics mathematics." (Paul Samuelson in Dore et al. 1989, 112.) See also Shubik 1992.

3. On the use of this phrase concerning von Neumann in a different context, see Heims 1980, 293 and Cini 1985.

gram in mathematics, sometimes called the "formalist" school of metamathematics.[4] According to Hilbert, one could not deduce mathematics from logic alone, as in the logicist program of Russell and Whitehead. Axiomatization was the key to both differentiation and guarantee of rigor in all mathematical disciplines, with each having its own characteristic axioms. The extent to which this was regarded the *sine qua non* of all thought can be gleaned from Hilbert's own words of 1922:

> The axiomatic method is indeed and remains the one suitable and indispensable aid to the spirit of every exact investigation no matter in what domain; it is logically unassailable and at the same time fruitful; it guarantees thereby complete freedom of investigation. To proceed axiomatically means in this sense nothing else than to think with knowledge of what one is about. While earlier without the axiomatic method one proceeded naively in that one believed in certain relationships as dogma, the axiomatic approach removes this naiveté and yet permits the advantages of belief. (Quoted in Kline 1980, 193)

In this stance, one should not regard mathematics as distilled factual knowledge (perhaps in the manner of Mill), but rather as the formal manipulation of abstract symbols leached of all intuitive reference or meaning. By tracing all statements back to the relevant axioms by means of relatively mechanical and impersonal proof procedures, one ensured that proof and rigor would be uncontentious and objective. In the 1920s, Hilbert and his students, among whom was numbered John von Neumann, developed the metamathematical program of checking for the consistency of existing formal systems, axiomatizing many systems along the way. Indeed, it is significant that the first presentation of the ideas in the 1928 game theory paper were delivered to Hilbert's Göttingen seminar on 7 December 1926.

This drive to axiomatize dominates the work of von Neumann in the 1920s. Two of Hilbert's enthusiasms, the formalization of set theory and the new quantum mechanics, were bequeathed as problems to his young protégé. Von Neumann's early attempt to specify a concise set of axioms for set theory (1961, 35–57), in the words of Stanislaw Ulam "seem[s] to realize Hilbert's goal of treating mathematics as a finite game. Here one can divine the germ of von Neumann's future interest in computing machines and the mechanization of proofs" (1958, 11–12). This ideal of Hilbert which suggested that somehow mathematics transcended mun-

4. The formalist program is discussed with great clarity in Kline 1980 (chap. 11).

dane interests and confusions remained with von Neumann long after the formalist program itself had been relinquished. In a late speech on "The Role of Mathematics in the Sciences and Society" in 1954, he practically echoed Hilbert's prejudices: "Mathematics furnishes something that is quite important, namely that it establishes certain standards of objectivity, certain standards of truth; and it is quite important that it appears to give a means to establish these standards rather independently of every thing else, rather independently of emotional, rather independently of moral, questions" (von Neumann 1963, 6:478). This faith in noncontextual impersonal rules behind or below the surface phenomena of human interaction, it should go without saying, is a necessary prerequisite for the very idea of game theory.

But over and above Hilbert's pregnant metaphor of mathematics as a "game," the youthful von Neumann found himself reacting to the earlier 1908 attempt at formalizing set theory by Ernst Zermelo. The avid avatar of axioms was familiar with Zermelo's formalist exercises, including his 1913 paper proving one of the first theorems on chess.[5] These influences would suggest a structural homeomorphism between the formalist program and the very idea of a game, with its play with chessmen of indifferent semiotic reference, within rules which guaranteed the integrity of play. Thus, in contrast to Ulam (1958, 7) who claims that von Neumann derived his inspiration from Borel, I think that is more likely that the formalist program lent itself rather readily to a notion of the axiomatization of games, and that von Neumann's distress at Fréchet's (1953) claim that Borel should be considered the inventor of game theory derives from this source rather than simply from wounded ego.[6]

While Hilbert's formalism was clearly one of the important inspirations for game theory, the other important source was the rise of quantum

5. The theorem, which is hardly earthshaking, was that "either white can force a win, or black can force a win, or both sides can force at least a draw." See the discussion of this theorem in Aumann 1989 (chap. 1). In Kuhn and Tucker 1958 (105) we learn that as early as 1927, von Neumann had corrected an error in the Zermelo chess paper in some correspondence with one Dr. König.

6. "I cannot resist the itch of answering this rather infantile attempt to attribute everything to Borel and to say that everything that can by no stretch of the imagination be attributed to him is wrong or previously known or trivial or irrelevant or all of these together" (von Neumann to Morgenstern, 2 January 1953, Box 15, OMP). Further evidence for the importance of the Hilbert formalist program in the conceptualization of game theory comes from the fact that another mathematical anticipator of some of its structures, Hugo Steinhaus (1960), was also a Hilbert student. On Steinhaus, see Leonard 1992.

mechanics in the 1920s.[7] It should be noted that Hilbert himself was intrigued by what his former student Max Born had been doing with integral equations and infinite bilinear forms, and in the mid-1920s turned his own hand to the rapidly developing theory of quantum mechanics (Mehra and Rechenberg 1982, 30ff). He put his star student, von Neumann, to work on this project as well, and the original research on the axiomatization of quantum mechanics and early game theory work were carried out simultaneously and both published in 1928 (von Neumann 1961, 1:104–33).

While any two fields of intellectual endeavor would not seem at first blush more alien to the tyro than these, there actually were some close similarities in their respective structures. The distinguishing characteristic of the "new" quantum mechanics (and some would still say its Mark of Cain) was its introduction of stochastic relations at the most fundamental level of the postulates of physical law, to such an extent that reconciliation with classical physical concepts was difficult, if not impossible. From the vantage point of von Neumann, one of the early charms of game theory is that it would have the same effect upon the theory of rationality. One of the main tenets of von Neumann's extensive writings on quantum mechanics was to demonstrate that the statistical character of the theory was not due to the ignorance of the observer about the exact quantum state, but rather that the system comprising both the observer and the observed phenomenon gives rise to the uncertainty relations even if one admits an exact state of the observer (Ulam 1958, 22). The 1928 game theory paper goes out of its way to make an analogous point: "Although in Section 1 chance was eliminated from games of strategy under consideration (by introducing expected values and eliminating 'draws'), it has now made a spontaneous reappearance. Even if the rules of the game do not contain any elements of 'hazard'—in specifying the rules of behavior for players it becomes imperative to reconsider the element of 'hazard' " (von Neumann 1959, 26). This

7. Thus this paper should be seen as an amendment to my theses on the relationships between physics and neoclassical economics (1989a and 1989b). In those earlier works, I had claimed that quantum mechanics had posed a difficult challenge to those who had become partisans of the nineteenth-century metaphor of utility as potential energy, and that the "resolution" of that challenge was to incorporate stochastic concepts in a harmless way into "econometrics." I would now like to claim that game theory was itself a second proposed *rapprochement* with the new physics; but that it also brought in train further difficulties for the core of the neoclassical program. The importance of quantum mechanics for von Neumann and Morgenstern has been noted in passing by Ingrao and Israel (1990, 186).

nodded toward the idea of a mixed strategy, so very important for the minimax proof, of which more anon. What is relevant for the present topic is that von Neumann regarded this appearance of intrinsic stochastic behavior as some sort of indication of a basic shared principle of the social and the natural, which only was reinforced by the metaphysical discussions that were the trademark of the Copenhagen Interpretation.[8]

Although the contested character of the stochastic was a major common denominator of quantum mechanics and game theory, the homeomorphism did not stop there. Both tended to resort to some of the same mathematical devices, in particular taking as a starting point Hilbert's theory of quadratic matrix forms (Mehra and Rechenberg 1982, 121–23). Both depended very heavily upon taking mathematical expectations, to such an extent that von Neumann signaled in the 1928 paper that he was not interested in dealing with objections to applying moral expectation to utility theory (von Neumann 1959, 16n). There were also other, more subtle ways in which von Neumann's treatment of quantum mechanics would reveal parallels to his later work in game theory. For instance, there was the lurking problem of intersubjectivity in his 1930s formalization of quantum mechanics:

> If one observer's "awareness" influences the physical event being observed, the question arises whether a second observer who becomes aware of the same physical event will agree in his observations with the first. To obtain intersubjective agreement, von Neumann offers only the possibility that the second observer can regard the first observer as an object, a measuring instrument, and by proper use will arrive [at the same results]. However, he has no mechanism for obtaining agreement between two observers, each of whom has an independent "consciousness." (Heims 1980, 135)

The analogous inability to permit the existence of two divergent rationalities in the theory of games would severely limit its applicability. Further, in his work on quantum measurement, von Neumann was forced to assume that the act of measurement was instantaneous, forestalling a true

8. A good introduction to the Copenhagen Interpretation of quantum mechanics is Gibbins 1987 (chaps. 4 and 5). Parenthetically, one of the reasons for von Neumann's unstinting hostility to the work of Paul Samuelson was probably the latter's loose usage of some Copenhagen terminology, such as the "correspondence principle" and "complementarity," in such a way as to guarantee the disdain of those conversant with the issues involved.

dynamic treatment. This restriction to the ideal static case, mainly dictated by the choice of mathematical tools, also looms large in the later work on game theory.[9]

Thus game theory became one more instance of how conceptions of the Natural feed back upon the definition of the Social (Mirowski 1989a); and the cybernetic conduit in this case was John von Neumann. The games/quantum mechanics nexus may have also played a facilitating role in the appearance of the 1944 volume, for the third party present at the tea at the Nassau Club on 1 February 1939, where von Neumann finally took notice of Oskar Morgenstern, was none other than Niels Bohr, who was holding forth (as usual) on the Copenhagen Interpretation and its metaphysical implications.

But, as they used to say in eighteenth-century novels, we get ahead of ourselves. Let us return briefly to the beginning of the 1930s, and ask, Why did von Neumann publish nothing more on game theory until the 1944 treatise? The chronology leading up to the collaboration with Morgenstern on TGEB is discussed in Leonard 1992, where we learn that von Neumann had resumed giving some lectures on games in the period 1937–40. We also know that von Neumann had individually written a sketchy mathematical summary of the theory of games, never to be published, including the first version of his stable-set solution concept but excluding any reference to applications, in October 1940.[10] The reconstruction of his motivations must necessarily assume a speculative character in this regard, since von Neumann never reflected back upon the lapse between the 1928 paper and TGEB in either published articles or unpublished correspondence. However, the origins of game theory in formalist metamathematics and quantum mechanics may point us in the

9. "In order to achieve this miracle [of mathematical structures of great generality] all the existing troubles had to be pushed into the *relation* between theory and fact, and had to be concealed. . . . Von Neumann's work is an example of this procedure. In order to arrive at a satisfactory proof of the expansion theorem in Hilbert space, von Neumann replaces the quasi-intuitive notions of Dirac and Bohr with his own. The theoretical relations between these notions are accessible to a more rigorous treatment. . . . It is different with their relation to experimental procedures. No measuring instruments can be specified for the great majority of observables, and where specification is possible it becomes necessary to modify well known and unrefuted laws in an arbitrary way or else to admit that some quite ordinary problems of quantum mechanics, such as the scattering problem, do not have a solution. Thus the theory becomes a veritable monster of rigor and precision while its relation to experience is more obscure than ever" (Feyerabend 1975, 64).

10. "Theory of Games" Part I: Analytical Foundations. Part II: Decomposition Theory. Part III: General Considerations about Solutions (von Neumann Papers, Box 23).

direction of an explanation of this lapse and his renewed (but temporary) enthusiasm for games in the period 1937–44.

One of the major turning points of von Neumann's intellectual life came on 7 September 1930, when Kurt Gödel announced his incompleteness proof. Von Neumann was visibly agitated by the announcement and drew Gödel aside to press him for further details (Shankar 1988, 77–78). On 20 November he wrote to Gödel to announce the discovery of a corollary, the unprovability of consistency, only to discover that Gödel had already arrived at the same result and had incorporated it into the text of his (now famous) paper. Once absorbed, it became apparent that Gödel's proofs had torpedoed the entire Hilbertian formalist program (Kline 1980, chap. 12; Shankar 1988). As Kline puts it, "Gödel's result dealt a death blow to comprehensive axiomatization" (1980, 263). By 1931, von Neumann was already lecturing on Gödel's proofs to seminars at Princeton (Shankar 1988, 48). But this was something more than another stunning mathematical accomplishment: it stole away at least half of von Neumann's mathematical *raison d'être*. He told numerous people at Princeton that after Gödel had published his paper, he did not bother to read another paper in symbolic logic (Kuhn and Tucker 1958, 108). The project of comprehensive axiomatization also lost its luster. In his famous lecture on "The Mathematician" he offered what was regarded as a poignant recantation of his youthful enthusiasms:

> The very concept of "absolute" mathematical rigor is not immutable. The variability of the concept of rigor shows that something else besides mathematical abstraction must enter into the makeup of mathematics. . . . My personal opinion, which is shared by many others, is that Gödel has shown that Hilbert's program is essentially hopeless. . . . I know myself how humiliatingly easily my own views regarding the absolute mathematical truth changed during this episode, and how they changed three times in succession. (von Neumann 1961, 1:4–6)

To let slip the epistemological anchor of absolute mathematical rigor in the very prime of his intellectual life could not have been a pleasant experience for such a one as John von Neumann. However, far from being scuttled by the experience, I believe it eventually goaded him into producing a substitute for the Hilbert program in what would now be called "artificial intelligence." In his essay "The Mathematician," he warns of the danger of mathematics becoming too inbred and rarified, losing its touch with the real-world problems that were the source of all profound innovations in mathematics. It seems that he was making oblique reference to his own opinion of the mathematical logic of his time, which had reached the vexed impasse of Gödel's proof. Due to the loss of the beacon of absolute rigor, the solution was to turn back to the "empirical" phenomenon for inspiration: in this particular instance, that meant attempting to model the brain directly in order to gain insight into the knotty conundrums of logic. One observes this in almost all the topics von Neumann took up in the last fifteen or so years of his life: probabilistic logics, the theory of automata, the design of computers, direct models of the neurons of the brain, and, of course, game theory. (The only topics that do not fit this pattern were various mathematical questions revolving around the development of the atomic bomb, such as gas dynamics, nonlinear dynamics, and meteorology. The inducement to work on these topics derived from von Neumann's growing involvement in the military establishment.) Hence, while game theory began as an extension of the Hilbert program, with some nudging from Morgenstern it slowly became one component of the anti-Hilbert reaction.

This change in the leopard's spots might seem a highly improbable thesis, if one did not observe the identical phenomenon happening with another prominent mathematical logician of the same period, Alan Turing. It is eerie to observe the extent to which Turing was von Neumann's intellectual doppelgänger. Turing also started out being attracted to the seeming paradoxes of probability theory, writing his thesis on the central limit theorem. As a natural extension, he was also fascinated by the new quantum mechanics and read von Neumann's text on the subject with great relish (Hodges 1983, 79). Turing's main claim to fame was in mathematical logic, where his paper on computable numbers in a sense fleshed out Gödel's incompleteness theorem. Yet upon confronting the paradox that the computable could give rise to the uncomputable, Alan Turing did not crumple in defeat, but rather in words that could apply equally to von Neumann, "Instead of trying to defeat determinism, he would try to account for the appearance of freedom" (Hodges 1983, 107–8). He then blazed the same trail as von Neumann, becoming his only real contemporary rival in pioneering computer design and the theory of automata. (He also became heavily involved in military projects, but that part of the story is irrelevant to our present concerns.) Most significantly, he also grew absorbed in an attempt to formalize chess based on minimax principles to the extent that it could be played by machine; this was, for all intents and purposes, game theory as conceptualized in Manchester as opposed to Princeton.

It may be useful to inspect this reflection of game theory in a mirror darkly, since it may help clarify some later philosophical demurrers concerning the program. Turing's machine grew out of an image in his work on computable numbers which formalized computation as fully determined by the following items: an input, a set of machine configurations or "states," and a set of instructions to produce an output. While Turing thought of the input and output as an infinite tape upon which could be printed ones and zeros, in fact it was better regarded as a table of behaviors which fully described the process. This led to the perfectly abstract Turing machine, which could imitate the behavior of any other "machine." The thrust of his early work was to demonstrate that there was no mechanical decision procedure to ascertain whether a given goal could be reached or not by standard mathematical procedures. Yet this did not belie the fact that real chess players, real mathematicians (and yes, real code-breakers) did make those kinds of decisions on a daily basis. Thus the issue was still open as to the extent to which actual practices regarding inference could be simulated by machines. While Turing's early work "had focused attention upon the limitations of mechanical processes, it was now that the underlying materialist stream of thought began to make itself more clear, less interested in what could not be done by machines, than in discovering what could. He had demolished the Hilbert programme, but he still exuded the Hilbert spirit of attack" (Hodges 1983, 212). Ignoring the complication that von Neumann never actually produced an impossibility theorem himself, the same could be said about his own reaction to the demise of the Hilbert program.

The parallel even extends to the fact that Alan Turing found his Oskar Morgenstern–style amanuensis in the guise of David Champernowne, who participated in formulating the chess-playing program using minimax decision procedures. But if the comparisons are so close, why didn't the Turing/Champernowne team develop game theory in its more narrow mathematical definition or extend its application in the direction of economic issues? While part of the answer must lie in Champernowne's lack of any profound involvement, pro or con, with the neoclassical program—a quirk of many technically proficient Cambridge economists of that era—most of the explanation must rest with Turing's own irony about his quest. As he wrote in a defense of his attempts to program computers to play chess,

> The reader might well ask why we bother to use these complicated and expensive machines in so trivial a pursuit as playing games. . . . We have already explained how hard all programming is to do, and how much difficulty is due to the incompetence of the machine at taking an overall view of the problem which it is analyzing. This particular point is brought out more clearly in playing games than in anything else. . . . Research into the techniques of programming . . . may in fact lead to important advances, and help in serious work in business and economics—perhaps, regrettably, even in the theory of war. We hope that mathematicians will continue to play draughts and chess, and to enjoy themselves as long as they can. (Turing quoted in Bowden 1953, 287–88)

In his comment on "the overall view," Turing reveals his discomfort with the minimax criterion, with its rather myopic and defensive conception of strategy, as well as his belief that parlor games only highlight this deficiency. It was simply too premature to claim that synthetic rationality had much to offer to economics as yet; a position never even entertained by the Princeton duo. Turing was not at all abashed about the overall objective of the program, which was to eventually simulate all aspects of human intelligence, as he asserted in his famous article in *Mind*. The only problem was that, as he admitted, the only chess player he had succeeded in getting the machine to mimic was himself: a rather sloppy strategist, and hardly the epitome of rationality (Bowden 1953, 245).

One implication of this history is that one way to understand game theory as it grew in significance for John von Neumann is to regard it as another variant of the Turing Machine; an attempt to plat the boundaries of formal rationality in reaction to the failure of the Hilbert program. As the Turing Machine is supposed to simulate the behavior of any other machine, game-theoretic strategic considerations are supposed to simulate the behavior of any opponent and therefore serve as a general theory of rationality. Both versions tend to display the same weaknesses as well; and this brings us to the quote that prefaces this paper.[11]

11. There was a third contemporary who began as a logician in this narrative, namely Ludwig Wittgenstein; yet his response to the upheaval of the 1930s was to endow "games" with an entirely different meaning. After his rejection of his own *Tractatus*, he came to believe that the mechanical formalization of logic would not solve any philosophical problems; as the first quote reveals, this practice should be thought of as just another language game (Wittgenstein 1975). It was his contention that it was unintelligible to speak of any machine as "following a rule," and that this encouraged a certain species of category error, where the cultural practices

Returning to von Neumann, we find that he seemed to be captivated by the image of the Turing Machine (Hodges 1983, 145). In his Sillman Lectures he explained, "The important result of Turing is that in this way the first machine can be caused to imitate the behavior of *any* other machine. . . . Generally speaking, anything that the first machine can do in any length of time and under the control of all possible order systems of any degree of complexity may now be done as if only 'elementary' actions—basic, uncompounded, primitive orders—were involved" (von Neumann 1958, 73). This goal of a modular, generic, mathematical, rule-governed rationality which could simulate any eventuality dominated von Neumann's efforts for the rest of his life. It was the image of the human brain as an electronic Turing Machine that informed game theory but which then came into its own when he realized that his interests in quantum mechanics, probability theory, and logic could be better reconciled by direct modeling of the neuron structure of the brain:

> [The brain] is a sort of a frequency-modulated system of signalling; intensities are translated into frequencies. . . . The message-system used in the nervous system, as described in the above, is of an essentially *statistical* character. . . . What matters are not the precise positions of definite markers, digits, but the statistical characteristics of their occurrence. . . . It is, therefore, perfectly plausible that certain (statistical) relationships between such trains of pulses should also transmit information. In this connection it is natural to think of various correlation-coefficients, and the like. (von Neumann 1958, 77–80)

It is significant that von Neumann's return to a concern with game theory roughly coincided with his first mention of the origins of this program. In an episode Aspray (1990, 178) calls important for von Neumann's intellectual development, he initiated a correspondence with the Hungarian physicist Rudolf Ortvay, which eventually stretched to approximately sixty letters, on the implications of quantum theory for the brain. In a letter dated 29 March 1939, von Neumann admits he had been thinking "mainly from last year" about the problem of the observer in quantum

of the programmer were mistaken for abstract and general principles somehow embodied in the machine. The machine never "interpreted" a rule that existed free of context; it was simply deployment of the choices of the programmer masquerading as disembodied logic. While Wittgenstein did not, to my knowledge, ever comment directly upon game theory, his numerous remarks upon mathematics and language games will provide an alternative reading of the history of games throughout this paper. It will be discussed in detail in my forthcoming paper "Machine Dreams."

mechanics as a "quasi-physiological concept." Ortvay raised the comparison between the brain and an electronic calculator; von Neumann responded by sending the draft manuscript on the theory of games. The fact that game theory and artificial intelligence tended to blur together in this incipient stage is illustrated by a response of Ortvay on 29 January 1941:

> I liked your paper [on the theory of games] very much at the time and it gave birth to the hope that, if I could succeed in drawing your attention to the problem of brain cells, you may succeed in formulating the problem. The problem seems to be this: The brain can be conceived as a network, having in its nodes the brain cells. These are interconnected in a way that each cell can receive from and transmit to several other cells impulses. . . . The actual state of cells conceived as numbered ones could characterize the state of the brain. There would be such a distribution corresponding to every spiritual state and the state would be relevant to every reaction, e.g. that which is the way of transmitting a stimulus to a nerve. (Quoted in Aspray 1990, 179)[12]

In this light another way to read von Neumann's project after Gödel's theorem is to see it as an attempt to displace the rational energetics of the nineteenth century (Mirowski 1989a) with a more up-to-date and mathematically rigorous version for the twentieth century. Instead of the older image of energy fields suffusing all human consciousness and being subject to deterministic constrained maximization, here we have trains of electronic impulses which depend in a fundamental way upon the stochastic character of quantum mechanics. Probabilistic logics meet automata theory, and the progeny is the Turing Machine made flesh. Game theory *per se* might not have continued to play a major role in this program, unless someone could come along and convince von Neumann that this very same machine-rationality dominated some distinct area of social life—say, for instance, the economy. That person was Oskar Morgenstern.

12. Unfortunately, the entire correspondence has never been translated from the Hungarian. On 16 February 1941, Ortvay even provides the metaphorical bridge to economics! "These days everybody is talking about organization, totality. Today's computing machines, automatic telephone exchanges, high-voltage equipment like cascade transformers as well as radio transmitting and receiving equipment, but also an industrial plant or an office are all technical examples of such organizations. I think there is a common element in all these which is capable of being axiomatized" (180).

Oppressed by the Limitations: Oskar Morgenstern (1902–1977)

Oskar Morgenstern was not your garden-variety orthodox neoclassical economist. Indeed, both admirers and detractors have entertained the proposition that he was not really an economist at all. For instance, how many people would admit that they were "deflected to soc.sc. by war (WWI); inflation and revolution in the streets, home difficulties, but *not* by deep intellectual attraction" ("General Notes" dated 4 August 1975, Box 4, OMP). Sometimes third parties see us more clearly than we see ourselves; and in one case, I believe that his correspondent Eve Burns summed up his past and future career as early as 1934: "I am very glad to hear that you have turned to the Zeitmoment work as I was beginning to get a little sad at the so very prominent negative turn in all your writings. [It seems you want to be] the first to prove that forecasting is NOT possible, the second apparently to show the LIMITATIONS of economic control, etc."[13] This predilection toward a delicate hypersensitivity regarding the limitations of our social actions, be it in the sphere of government control, or forecasting, or data quality, or the economics profession, or of rationality itself, is the leitmotif of all of Morgenstern's thought. He recognized this in himself obliquely by stressing retrospectively his studies in logic with Alfred Whitehead and the Vienna Circle and by claiming that he was actually a philosopher *manqué:* "Methodology has occupied my mind, springing from the conviction that the social sciences are in greatest need to find their philosophical basis which they are sadly lacking. But I have always felt that concrete work in economics is necessary in order to speak with confidence about method" (Brief autobiography, dated 1975, Box 9, OMP).

Morgenstern was lucky enough to be remembered for some of this "concrete work," although it should be said that throughout his life the concrete work was mostly done in collaboration, and that the portion of that work that can be attributed to himself in the mathematical mode was slim to nonexistent. This was more due to deficiencies in training rather than inclination, since one other major theme of his intellectual life was his spectator's fascination with the role and uses of mathematical

13. Eve Burns to Morgenstern, 25 January 1934; Box 2, OMP. Eve Burns was an early advocate of statistical techniques in economic empiricism, a professor at Columbia University, and one of the reviewers of Morgenstern's first book (Burns 1929). There is an extensive correspondence between Burns and Morgenstern from the 1920s and early 1930s, in which he expresses his private feelings much more openly than in any other extant correspondence.

formalisms. This rather set him apart from his Viennese colleagues in economics, who were under the sway of Mises's hostility to mathematics. Yet even here, this fascination was motivated by the connoisseur's appreciation of limitations, since "one could say that our science, especially logic and mathematics, is a measure of the inadequacy of the human mind" (quoted in Schotter 1976, 394).

In most other respects, his early interests were by and large those of a segment of the community of Austrian business cycle theorists consisting of Wieser and his mentor, Hans Mayer (Rellstab 1992). He seldom missed any opportunities to cite Menger, Böhm-Bawerk, or Wieser as progenitors of neoclassical ideas and was absorbed by problems of temporal phenomena and economic dynamics. He was skeptical of prevalent notions of economic "equilibrium" but tended to give it a particularly contemporary spin by stressing the existence of "indeterminacy" in economic life.[14] The recognition of the chaotic led quite naturally to a concern with the conceptualization of risk and the place of probability theory in economics. One of his most famous papers of the 1930s insisted that perfect foresight, if such a thing could ever exist, would *not* banish risk since "always there is exhibited an endless chain of reciprocally conjectural reactions and counter-reactions. This chain can never be broken by an act of knowledge but always only through an arbitrary act—a resolution. . . . Unlimited foresight and economic equilibrium are thus irreconcilable with one another."[15] The little parable that Morgenstern never missed an opportunity to repeat in order to illustrate this point, beginning with his 1928 book *Wirtschaftsprognose* and ending with his valedictory (1976), was from the Conan Doyle story "The Final Problem" where Sherlock Holmes and Moriarty try to outwit each other in taking (or not taking) various trains to reach the Continent. Herein lies the germ of the idea of interdependent strategies which both von Neumann and Morgenstern regarded as the defining characteristic of game theory.

Nevertheless, at the outset of his career, the young Morgenstern was

14. For instance, in his *Limits to Economics* we find: "It may be generally accepted that the strength of economic 'power' may be measured by the amount of indeterminacy which it introduces into the economic process. . . . The economic process is characterised by a much higher degree of *indeterminacy* than had previously been supposed" (Morgenstern 1937, 66, 104). One can observe in this both a basis for common ground, given von Neumann's interest in quantum mechanics, and the foreshadowing of the subsequent hostility to the search for a unique *optimum optimorum* solution concept in game theory.

15. The quotation is from an English translation of Morgenstern 1935 (11–12), Box 5, OMP.

not so very certain that mathematics could or would rectify any of these knotty problems. One of his earliest letters to Eve Burns chides her statistical enthusiasms: "Have you well considered the fact that the problem of 'verification' of economic theory by means of statistics is a very difficult methodological question at the same time? Here we are also studying mathematics although we grow more and more skeptical about its use, the main objections coming from the fact that time cannot be dealt with adequately" (undated letter [probably late 1927], Box 1, OMP). Eve Burns was not one to take such a rebuke lying down, tendering a spirited defense, which prompted Oskar to once more state his position:

> [In] my lecture from Vienna on Mitchell and Pigou . . . I deal extensively with the question of quantitative and qualitative methods and with the problem of statistification of economic theory etc. . . . You have my ideas on the subject stated there neatly and it is no good that I should repeat myself now only much less complete. This paper has stimulated me to continue these ideas and to take up some older notes from Harvard concerning the invalidity of the concept of equilibrium etc. . . . All this makes me still more suspicious of the mathematical method, but nevertheless I am trying to get familiar with it as much as I can. . . . You say that you study science. This is very laudable, but again I want to warn you not to forget, that the method and aims of our social sciences are very much different from the natural sciences" (2 March 1928, Box 1)

Thus the early Oskar Morgenstern looked more or less like a typical Austrian economist of the fourth generation, intent upon projecting the archtypical Austrian concerns about knowledge and time and the meaning of science against a backdrop of neoclassical theory. "Sometime, if I can, I should try to bring about a reconciliation between the mathematical school of Lausanne and other groups, based on a careful analysis of methodological assumptions. That means, of course, that one should have to do away with such things as ophelimité ponderée etc" (Letter to Allyn Young, 22 March 1928, Box 1, OMP). This presumption that methodological fiat could solve substantive problems in economic theory assumed the format of a belief that formal logic and metamathematics had something to offer economics. To a lesser degree than the Hilbertian formalist program, this tended more to privilege Russell's and Whitehead's *Principia* (due to study with Whitehead at Harvard) and Wittgenstein's *Tractatus* (which was enjoying the height of its enthusiasm among Schlick's Vienna Circle).

There is no evidence that Morgenstern actually became embroiled in such issues as the set-theoretic paradoxes or the validity of logical atomism; and the lessons actually drawn from logic for economics rarely transcended the rather prosaic statement, "The principle of freedom from inconsistency of economic policy is the only scientific-economic principle which can be formulated without passing value judgments" (Morgenstern 1937, 53–54), or the suggestion that the Russellian theory of types had something to offer, a promissory note never actually to be cashed out (Schotter 1976, 395). He did wax enthusiastic over Hilbert's program of axiomatization in 1936 and expressed the opinion that it would be fruitful in economics; but again, enthusiasm never gave way to actual accomplishment. Nevertheless, this interest in logic did lead to a presentation of the Holmes/Moriarty problem to Karl Menger's mathematical colloquium, which in turn created the occasion for one Eduard Čech to approach Morgenstern after the talk and suggest that one John von Neumann had already provided a formalism in 1928 to deal with the issue (Morgenstern 1976, 806–7). This experience prepared the ground for Morgenstern to search out von Neumann in the hope that he would lend his support to the required methodological critique of economics.

What is curious is that, although they were both Viennese intellectuals, the fact that the meeting took place at all was largely due to accident. Having gotten his Dr. Rer. Pol. at the University of Vienna in 1925, Morgenstern was one of the chosen few to be the object of lavish support from the Rockefeller Foundation in the period in which it handpicked European economists who would redirect economics into a more "scientific" modality (Craver 1986b). From 1925 to 1928 he traveled widely, including stints at the LSE, Columbia, and Harvard, without any strict requirements to actually produce work on schedule.[16] Upon his return to the University of Vienna, he was named head of the Rockefeller-funded Institut für Konjunkturforschung after Hayek departed to accept the chair at the LSE. In January 1938, he departed on another Carnegie-funded jaunt and so just happened to be in the United States when Hitler occu-

16. His progress report to the Rockefeller fund dated November 1926 revealed the rather easy-going nature of funding agencies in those days: "The time until June 9th I spent in New York, at Columbia University. I got close contact with Professors Mitchell and H. L. Moore and with Professor F. A. Fetter from Princeton. In the meantime my studies of business cycles—the main topic of my studies—took more and more shape and now at present have a very definite form. . . . I am, however, unable, at the moment to give any dependable information about the prospects of my studies of business cycles. My study is at a very large extend methodological and studies of this kind seldom make a smooth progress and one never knows about complications that may arise" (Box 22, OMP).

pied Austria in March of that year. When warned by friends that his return would not be regarded favorably, he abruptly decided to remain in the United States. With some difficulty but aided by Frank Fetter, he found a temporary half-time post at Princeton, with the Rockefeller Foundation picking up the rest of the tab. Thus the stage was finally though inadvertently set for a meeting between von Neumann and Morgenstern, although it was by no means a foregone conclusion that it would have momentous consequences.

The Nature of the Common Ground

Oskar Morgenstern once jotted down the following question, more in gratitude than in wonder: "Why does J. take so much pleasure to discuss mathematics with me *way* beyond what he told people like Wald?" (Box 23, OMP).[17] The question does need to be asked, given Morgenstern's tenuous grasp of the technical issues involved and the somewhat divergent sets of interests outlined above. What was it that drew such unlikely fellows together and eventually convinced the superhumanly busy von Neumann to collaborate upon the composition of TGEB?

There was, first and foremost, the phenomenon of being Austrian émigrés together in a strange land. (This was also a significant factor in the subsequent development of game theory in the defense community [Mirowski 1991].) When working together, they always spoke German, even though they were writing the text in English. Their circle of friends consisted mostly of fellow émigrés such as Gödel, Wigner, and Weyl. Von Neumann had a notoriously raunchy sense of humor, and Morgenstern was one of the few intellectuals in the rarefied atmosphere of Princeton to encourage it.

Nevertheless, such personal considerations would not be sufficient to initiate the long and arduous task of composing 600-plus pages amid the hustle and bustle of World War II; and they were not. The chronology is discussed in Leonard 1992. In 1939 and 1940 von Neumann had extended his mathematical work on games, but it was rapidly becoming eclipsed by events and the growing conviction that the axiomatization of the brain was the ultimate terminus of a comprehensive theory of logic. A perusal of the unpublished 1940 manuscript "Theory of Games" should have raised the issue of the intended audience. There were no mathemati-

17. In a notebook entitled "Notes on Johnny's opinions about I-O, θ, computat., etc."

cians actively researching these problems, and the manuscript itself gave absolutely no room to applications and no quarter to the mathematically unsophisticated. War work was increasingly claiming more of von Neumann's time, and the correspondence with Ortvay was nudging him in the direction of the computer. However, starting in April 1940, Morgenstern began to raise the possibility of potential connections between economics and the theory of games.

Morgenstern's diary entry of 9 April 1940 reads, "N. read Zeitmoment and found it very interesting and full of difficult problems. He said it was very concise and exact. I ought to put all this—risk, foresight, revenue, etc—together in a book." On 19 August 1940, after a discussion with von Neumann about his essay, he wrote, "The combinatorial things are complicated and not very clear." On 31 August they discussed the connections to polyvalent logic and quantum theory, and on 20 September von Neumann "finds that one of [Morgenstern's] functions bears a striking resemblance to an important function in thermodynamics. It also provides great analogies to higher and lower energies, etc." On 19 October there is the entry "Lunch with vN. I told him about some criticisms of the theory of indifference and he agreed with me. He is also convinced that marginal utility is not coherent. I should work all that into the Hicks review. It would be good if he were to write a mathematical supplement." On 8 November: "Yesterday I showed him the contract curve. It has a bearing on his games, namely because it also matters who goes first." On 4 February, von Neumann read the proofs of Morgenstern 1941 and made suggestions for revisions. Until July, Morgenstern made various comments to the effect that the mathematics was giving him trouble and began to hope for a collaboration with von Neumann, one that would make so many things so much clearer.[18]

What was it that finally turned "Johnny" around, and brought about a convergence of interests? Perhaps not so strangely, it seems to have been shared attitudes (or Morgenstern's adopted attitudes) regarding von Neumann's driving motivations, quantum mechanics and logic. At the

18. Box 4, OMP Diaries. All translations from the German courtesy of Pamela Cook. All subsequent entries from Morgenstern's diaries will be listed in the text as (date, OMPD). Morgenstern also mentions the quantum connection in his valedictory (1976) and in a note dated 4 August 1952 entitled "Uncertainty in Social Behavior" (Box 22, OMP), which sums up the contention in the text: "(1) Distinct: (a) Mech, classical & stat & (b) Quant. Mech. (2) Similar in Soc. Sciences?? Can all uncertainty be reduced as theoretically in (a) or exists some sort of (b)? (3) Remember that each field may require its own logic—as (b) does."

tea that brought Morgenstern to von Neumann's notice, Bohr was propounding the thesis that "physics is too often taken as a model for the philosophy of science" (15 February 1939, OMPD). It was this question of whether or not physics should or could provide the paradigm for social theory and philosophy of logic that fascinated the duo, although it must be said that von Neumann fairly consistently favored the position that modern (i.e., post-quantum) physics should be the template, whereas Morgenstern generally went looking for inspiration outside of physics. Nonetheless, Morgenstern was easily swept up in von Neumann's enthusiasms,[19] reporting in September 1940 on analogies with thermodynamics and commenting on a January 1941 discussion of von Neumann's work on quantum logic: "The nice, comfortable arrangement of science into logical and empirical compartments fails. That is immense. Everything comes from quantum mechanics" (21 January 1941, OMPD). Even when they were nearing the end of their collaboration, Morgenstern never lost sight of this conceptual gyroscope:

> I pointed out the problem of [max.-anschauung] of games. We discussed this and that resulted in a further paragraph of quite noteworthy content. Now there is still a passage to be written about maximum gain and maximum loss. Then came the individual examples. It is becoming clearer to me that the whole thing is like quantum mechanics. It will only last if that is absorbed. We both know that." (23 November 1942, OMPD)

The effect of this awareness was not a steady stream of analogies with the Schrödinger wave equation or anything else of the sort, but rather an unremitting stream of scorn expressed in the text of TGEB toward neoclassical economic theory because it was an imitation of outdated physics (von Neumann and Morgenstern 1964, 6, 45, 147). This also showed up at that time in von Neumann's refusal to review Paul Samuelson's *Foundations*, due to the fact that "one would think the book about contemporary with the time of Newton,"[20] and that Samuelson "has murky ideas about stability. He is no mathematician and one should not ascribe the analysis to him. And even in 30 years he won't absorb game theory" (8 October 1947, OMPD). One of the things that set

19. This is also suggested by Ingrao and Israel (1990, 409).
20. Note dated 15 December 1972 (Box 8, OMP).

the duo apart from most economist contemporaries is the clarity with which they understood that neoclassical economic theory was little more than a bowdlerization of nineteenth-century energy physics, combined with a resolve to do something about it.[21] Nevertheless, it appears that their respective disaffections were rooted in sharply divergent world-views. For von Neumann, it was the inept and rudimentary character of the mathematics that gave offense; he was rather quick off the mark to search for new economic analogies in modern physics, as we have witnessed. Morgenstern, on the other hand, seemed to retain some of his earlier conviction that the social sciences were intrinsically different from the natural sciences—an Austrian tenet of hallowed pedigree (Mirowski 1989a, 261, 355)—and that this implied the necessity for a new departure in formalization.

The quest for a more correct physical analogy did not simply harmonize the duo's skepticism about orthodox neoclassical practices; given the special context of the new physics of the 1910s and 1920s it also instructed them to concentrate their search in a few well-defined areas. The most obvious common ground was the shared conviction that probability theory must be comprehensively incorporated into economic theory at a fundamental level. For both von Neumann and Morgenstern, this happened where mixed strategies became indispensable for guaranteeing the existence of a minimax point. Only later did Morgenstern insist upon the incorporation of utility to denominate the payoffs, which in turn led to the von Neumann/Morgenstern expected utility, the subordinate role of which was openly admitted in the text: "We have practically defined numerical utility as being that for which the calculus of mathematical expectations is legitimate" (von Neumann and Morgenstern 1964, 28). Von Neumann's attraction to the intrinsic stochastic character of games has been documented above; it should be noted that Morgenstern partook of this enthusiasm as well. In some unpublished notes he wrote, "The *combinatorial* character of the theory of games is one of its more out-

21. The actual historical narrative of the primitive appropriation of the physics is provided in Mirowski 1989a. Morgenstern went out of his way to make this point in his numerous popularizations of game theory, such as this: "The notion of general economic equilibrium . . . is derived from mechanics, but mechanics is not the model to account for the phenomena. . . . This means asking whether another model can be found to replace the ideas derived from mechanics and their mathematical treatment which necessarily followed the same lines. It appears that games of strategy offer the desired analogue and that the theory of games provides the solution" (1949, 296–97).

standing features. This has very deep lying significance which is difficult to grasp" (Box 15, OMP).[22]

Another shared interest was the need they both felt to revise and refurbish the notion of generalized interdependence which has often been claimed as one of the virtues of the Walrasian research program but never actually realized in their mathematics (Ingrao and Israel 1990). Von Neumann was predisposed to accept this tenet by the Copenhagen interpretation of quantum mechanics; whereas it had been an indictment of the Lausanne school since Morgenstern's early career. Evidence of complementary interests in this regard can be derived from the paper on "Maxims of Behavior," which von Neumann encouraged Morgenstern to write in May 1941, just prior to the decision to collaborate: in stressing the distinction between individual plans that did and did not require taking the plans of others into account, it insisted upon a theoretical distinction between subjective and objective rationality not present in orthodox theory.[23] While their subsequent joint text stated their dissatisfaction with the Lausanne school's stifling the insights of interdependence by means of too many restrictive assumptions (1964, 15), they did not explicitly state the criteria by which they chose the assumed invariants which they themselves imposed upon the analysis.

Finally, the governing analytical presupposition that bound their programs together was a deep commitment to methodological individualism. "We believe that it is necessary to know as much as possible about the behavior of the individual and about the simplest forms of exchange" (1964, 7). Knowing about individuals is one thing, but insisting that the

22. Notes dated 25 December 1947. The notes continue, "(1) We note that comb. is at the very bottom of the foundations of math.—When something is settled combinatorially, it is really decided. Hence the fundamental nature of Gödel's work in math. logic. (2) On the other hand, the applic. of combinatorial methods is a sign that special math. techniques based upon it, are lacking (such as calculus, etc.). (3) Tendency to combinatorial reasoning in *physics,* e.g. Pauli's exclusion principle is of this nature and much of present quantum mechanics. [Discreteness as in Quantum Mechanics.] Formerly it was the idea of (time-space) continuum (relativity theory, Einstein . . .) which also has a strong direct, intuitive appeal.—Not so combin. (4) There is no doubt (Weyl) that we do not understand the deeper properties of combinatorics, of permutations etc. Math. philos. has little to say about the field that is really relegated to a subsidiary position there—quite wrongly. Weyl: Comb. is a thing for the younger generation."

23. "Maxims of Behavior," "Subject files" 1938–70, OMP; n.d. The following in particular must have warmed von Neumann's heart: "Maxims where no quantitative factor is made explicit but where it nevertheless exists are always restricted axioms" (12), i.e. plans that require taking into account whether others are following the same plan.

entire theory be reduced to the actions of atomic individuals is another; and the fact that this tenet went unchallenged for both von Neumann and Morgenstern was part of what made their otherwise scathing critique still acceptable for many neoclassicals. Morgenstern's individualistic bias derived from his Austrian background and rendered him unsympathetic to the wholist approach of the German Historicist school or the American Institutionalists (see his unpublished 1931 "Observations on the Problem of the American Institutionalists," OMP). For von Neumann, the quest to model people as little Turing Machines dictated that both consciousness and intersubjectivity were not central to the understanding of strategic thought; the only hope for a theory that could rival physics in its ability to isolate the law-governed character of social life was to ground it in the natural characteristics of the individual. As he once said, "It is just as foolish to complain that people are selfish and treacherous as it is to complain that the magnetic field does not increase unless the electrical field has a curl. Both are laws of nature" (quoted in Heims 1980, 327).

Contrapuntal Voices in the Theory of Games

To make a list of what brought them together is not to suggest that the end product, the theory of games, was somehow a smooth homogenization of all their concerns. By this I do not mean to say that von Neumann and Morgenstern ever were explicitly at odds about game theory. Indeed, whenever there was the merest whiff of disagreement, I have found that Morgenstern would invariably defer to von Neumann. Nevertheless, reading the early Morgenstern presents one with a whole series of philosophical concerns that were never addressed or pushed to the margins of discourse in TGEB. As for von Neumann, his patience with the economics community as a potential seedbed for the cultivation of the theory of games was very rapidly exhausted (Mirowski 1991); and one can make the case that there were good reasons on both sides to deem the marriage premature and the offspring stillborn. Hence the "theory of games" circa 1944 was neither a perfectly coherent nor a self-sustaining construct, even after 600-plus pages of novel and imaginative formalisms.

The evidence for von Neumann's dissatisfaction with TGEB comes from two sources: first, that the rejection of the message of TGEB by a broad range of economists led von Neumann to create a context for game-theoretic research in the burgeoning community of defense analysts (Leonard 1992, Mirowski 1991); and second, some scattered comments he made toward the end of his life. Since the first set of events has been dealt with elsewhere I shall concentrate here on the second set. One subset of those comments assumed the cast of a certain dismay that economists had missed a major premise of the theory of games, namely, the very indeterminacy of the outcomes. As Martin Shubik (1992) reports,

> It is my belief that von Neumann was even more committed than Morgenstern to the idea of a solution as a set of imputations. He felt that it was premature to consider solutions which picked out a single point and he did not like noncooperative equilibrium solutions. In a personal conversation with von Neumann (on the train from New York to Princeton in 1952), I recall suggesting that I thought that Nash's noncooperative equilibrium solution theory might be of considerable value in applications to economics. He indicated he did not particularly like the Nash solution and that a cooperative theory made more social sense.

But beyond some skepticism about this or that particular solution concept, it seems that toward the end of his life von Neumann had come to the conclusion that his foray into economic theory was not really a fruitful part of his program of the formalization of machine rationality. In other words, there was nothing wrong with game theory *per se,* or with his work on the computer or logical automata or modeling the neuron; it was rather that the common ground between these and economic discourse was too tenuous and underdeveloped. The prime text in this regard is a talk he delivered on 12 December 1955 entitled "The Impact of Recent Developments in Science on the Economy and on Economics" (von Neumann 1963, vol. 6). One has only to read between the lines to plumb the extent of self-reference in his opening statement that "there have been developed, especially in the last decade, theories of decision-making—the first step in its mechanization. However, the indications are that in this area the best that mechanization will do for a long time is to supply mechanical aids for decision-making while the process itself will remain human. The human intellect has many qualities for which no automatic approximation exists" (100). The prognosis was not to relinquish the quest for machine rationality, but rather to cut the ties with the economists: that is, to relinquish the second half of the title of TGEB. "For trying out new methods in these areas, one may use simpler problems than economic decision-making. So far, the best examples of this have been achieved in military matters" (100–101). This, of course, was an oblique reference to his success by 1955 in transplanting the active community of game theory researchers into RAND and the military. Finally, he made clear that his perceptions of the failure of the program in economics had nothing to do with the more conventional reasons tendered for why economics could not attain the status of physics.

> With regard to the application of the scientific method in economics, it is important to see which difficulties are real and which are only apparent. It is frequently said that the economy is not predictable by rigorous scientific analysis, because one cannot experiment freely. One should remember . . . astronomy is the one in which you can least experiment. . . . It is also frequently said that in economics one can never get a statistical sample large enough to build on. . . . However, if one analyzes this carefully, one realizes that in scientific research as well, there is always some heterogeneity in the material and one can never be quite sure whether the heterogeneity is essential. . . . What seems to be exceedingly difficult in economics is the definition of categories. (101)

Although he did not elaborate upon the conceptual problem in this talk, it is possible to construct a list of issues that appeared to block the program of machine rationality in economics for von Neumann. These fall under the rubrics of the existence and justification of the minimax solution; the problem of a theory of rule structures; the implausible reduction of the nonconstant-sum game to the zero-sum game; and the ubiquitousness of transferable utility. The attribute they all share is the seemingly stubborn inability of the formalism to map into what von Neumann would allow were important "categories" of the economic phenomenon.

It is interesting to see the way that the minimax is practically deconstructed as a solution concept in the course of the elaboration of TGEB. The real justification of the minimax is that von Neumann wanted to build upon the foundations of his 1928 paper, and thus, of course, upon his insights gleaned from quantum mechanics. But since TGEB is nominally about the economy, the justification is initially framed in the format of defining skillful play as choice of a strategy that will ensure a realized value of the game greater than or equal to some constant (TGEB, 107). Upon embarking upon the two-person zero-sum game, we are invited to see that the interests of the players are diametrically opposed (98). From there we are swept along to the proof that the existence of a saddlepoint implies that both players' minimax strategies coincide; the argument is clinched on page 128 where it is asserted that "in a zero-sum two-person game the rationality [i.e., the minimax strategy] of the opponent can be assumed, because the irrationality of his opponent can never harm a player." But in common parlance, this would not be regarded so much as skillful play as the neutralization of the significance of the opponent.

The authors then endeavor to preserve the minimax solution in the *n*-person framework by dividing the problem into two phases where there is an initial choice of coalitions, each indexed by its unique "value," and only then reducing the game to its two-person counterpart of the coalition playing "everyone else" (220). Yet the authors discovered that this entire structure rested upon the assumption of the zero-sum payoffs: no action could be allowed to alter the global magnitude of the payoffs. In retrospect we see that this is simply the imposition of a conservation principle, a common theme in the history of Western theories of economic value (Mirowski 1989a). But here the trouble really began, because what might be regarded as an expendable assumption from a mathematical vantage point was critical for the very "rationality" of the minimax solution from an economic point of view. Uneasiness over the inapplicability of the zero-sum assumption to the economy virtually pervades TGEB (147–48, 224, 290, 539–41).

The attempt to evade these problems led to the proposal of the stable-set solution concept later in TGEB; however, it was this provisional tactic which I believe offended von Neumann's sense of mathematical aesthetics. For instance, the inability to produce clear-cut results on existence betokened something was amiss; somewhat later von Neumann attributed this to the character of the intended subject matter. In some remarks at a conference at Princeton on 1 February 1955,

> von Neumann pointed out that the enormous variety of solutions which may obtain for n-person games was not surprising in view of the correspondingly enormous variety of observed stable social structures; many differing conventions can endure, existing today for no better reason than that they were here yesterday. It is therefore still of primary importance to settle the general question of the existence of a solution of any n-person game. Another gap in present knowledge pertains to formation of the equilibria constituting the solution of an

> n-person game; even the beginning of a dynamic theory is lacking. (Quoted in Kuhn and Tucker 1958, 103)

It seems clear in retrospect that there was significant confusion over whether the phenomenon was driving the mathematical results or vice versa; after all, processes with a substantial component of hysteresis would not normally be candidates for portraits of machine rationality. In part, this lack of coherence was highlighted after von Neumann's death by the William Lucas counterexample of a ten-person game which had no stable-set solution (Lucas 1969). While this demonstrated that part of the problem lay in the mathematics, von Neumann himself preferred to attribute it to the intractable character of the phenomenon itself, namely, the economy (if perhaps not privately to the economists).

But there was another Tar Baby concealed in this recalcitrant phenomenon, the "economy." There was a second crucial invariant hidden in the mathematics: not only were payoffs globally conserved, but so were the "rules" of the game. Indeed, one of the reasons that the minimax could so completely neutralize the problem of the rationality of the Other was that it never occurred to the Other to violate the bounds of the preset rules. Such an option never enters the "mind" of the machine; and good breeding prevented its appearance at the poker tables of Los Alamos and Princeton; but in the economy, well, that was a different matter. Rules involved interpretation, and to escape the infinite regress, "there seems to be no escape from the necessity of considering agreements concluded outside the game" (TGEB, 223). This in turn opened the Pandora's Box of ethics (263n) and, worse, problems of interpretation of the past and the future in a formalism which, the authors kept insisting, was strictly static. Here it was the *disanalogies* with physics which potentially threatened to wreck the entire formalism (Mirowski 1986). Von Neumann's response was similar to that in his formalization of quantum mechanics: just expand the formal character of the "space" in which the model would reside.

> Von Neumann then outlined the program of a new approach to the cooperative game by means of some (not yet constructed) theory of the *rules* of games. A class of *admissable extensions* of the rules of the non-cooperative game would be defined to cover communication, negotiation, side payments, etc. It should be possible to determine when one admissable extension was "stronger" than another, and the

> game correspondingly more cooperative. The goal would be to find a "maximal" extension of rules: A set of rules such that the non-cooperative solutions for that game do not change under any stronger set. If this were possible, it would seem the ideal way in which to solve the completely cooperative game. (In Kuhn and Tucker 1958, 103)

Although this suggestion dates from 1955, and therefore too late for von Neumann really to do anything about it, I do not think it was a serious proposal. The space of all configurations of rules could not be enumerated in such a way that all their relevant consequences could be formalized and the magic wand of constrained optimization waved over them. Interpretation could not be first exiled, only to be invited back in later, a more tamed and chastened guest of the Rational Machine. I believe one can still observe the fruits of this impasse in the modern proliferation of solution concepts in game theory.

As if these fundamental flaws of the project were not devastating enough, the theory of games required further ad hoc adjustments to render it translatable into economic discourse: there were, for instance, the requirement of conjuring up a "fictitious player" in order to effect the reduction of non-zero-sum games to their zero-sum counterparts (505–6); the bald assumption of the choice of the point of "maximum social benefit" in order to identify an equilibrium point (513); and banishment of any motive of one player to punish another (539–41). How this was supposed to be an improvement over the Walrasian orthodoxy was anyone's guess; although the similarities certainly encouraged the revanchist movement of the 1950s and 1960s to reabsorb game theory into the orthodox neoclassical research program. Finally, there was the requirement of the reintroduction of cardinal and interpersonally comparable and transferable utility especially in the context of the non-zero-sum game. This gave von Neumann some discomfort,[24] although I believe he basically never totally gave up his original tendency to think of payoffs in terms of money (TGEB, 249n).

When all was said and done, what was the specific economic content that was gleaned after the extended formal preliminaries in TGEB? A model of a two-person single-bid market where the "result" was that the final outcome must fall between the reservation price of the seller and the maximum value for the buyer, where the region of indeterminacy

24. See, for instance, his letter to Morgenstern dated 16 October 1942, in Box 23, OMP.

was even larger than in the original paradigm of Böhm-Bawerk's "marginal pairs" (556–64). Surely the elephant had strained embarrassingly mightily to bring forth a pathetic mouse.

One of the ironies of the situation was the extent to which Morgenstern's original concerns were trampled and mangled in the final text of TGEB. The man who wrote in 1935 that "it is clear that a theory of economic equilibrium which 'explains' only a *static situation, which is given as unalterable* and which, because of this basic assumption, is completely unable to say anything about the economy when a variation occurs, is utterly unimportant from a scientific point of view. It would hardly deserve the names of theory and science" (in Schotter 1976, 180–81), now had to continually insert apologies into the text for the static character of the model. It must have required extreme deference on the part of Morgenstern to conduct such an about-face on the basis that "Johnny says we should wait 300 years for an applicable dynamic theory or 100 years if one is exceptionally optimistic."[25] Or how must it have felt to become associated with the Sherlock Holmes/Moriarty anecdote as a trademark and to write in 1941 that "observed acts of behavior allow an indefinite number of interpretations regarding the plans from which they are assumed to have sprung" (381), only to participate in the construction of a model where all processes of interpretation are baldly assumed away by the solution concept?[26] Or to spend your life discussing the limits of rationality, only to become the advocate of a model where choice of a strategy must presume perfect foresight and understanding, and choice of a mixed strategy must incongruously associate rational choice with the flipping of a coin? It seems that very little of the characteristic Morgenstern ideas were preserved in TGEB.

If that was the case, why were there no complaints, either in the form of spats during the composition, or in the form of second thoughts after publication? The only, inadequate, answer is that Morgenstern was so in awe of "Johnny" and so unaware of all the implications of the mathematics that the only way this tension ever surfaced was in the curious contrapuntal voice in which TGEB is conducted. The text itself points out that many of the arguments are carried out in a contradictory manner:

25. Notebook entry dated 23 November 1947 (Box 15, OMP).

26. "In the minimax strategy of a zero-sum game—most strikingly so with randomized choice—one's whole objective is to avoid any meeting of minds, even an inadvertent one" (Schelling 1960, 96).

quasi-dynamic arguments often rest uneasily beside the static mathematics; coalition formation is treated as depending only on payoffs in the mathematics but also upon external influences in the text (263); the zero-sum condition is openly admitted to be in conflict with the value-enhancing character of the market (540); poor imitations of physics are disparaged in favor of other imitations of physics which may or may not fare better; and so on. Von Neumann, in his turn, was ambivalent about the exact isomorphism of the economic interpretations of the model, and so apparently let these very unrigorous tergiversations stand in the text. The end product was a book curiously at odds with itself. Such a book could hardly have been expected to have set the economic world on fire; and the truth of the matter is that in its initial phases, it did not.

References

Aspray, William. 1990. *John von Neumann and the Origins of Modern Computing*. Cambridge, Mass.: MIT Press.

Aspray, W., and A. Burks, eds. 1987. *Papers of John von Neumann on Computing and Computer Theory*. Cambridge, Mass.: MIT Press.

Aumann, Robert. 1985. What is Game Theory Trying to Accomplish? In *Frontiers of Economics*, edited by K. Arrow and S. Honkapohja. Oxford: Basil Blackwell.

———. 1989. *Lectures on Game Theory*. Boulder, Colo.: Westview.

Bianchi, Marina, and Hervé Moulin. 1991. Strategic Interaction in Economics. In *Appraising Economic Theories*, edited by Mark Blaug and Neil de Marchi. Aldershot: Edward Elgar.

Binmore, Ken. 1987–88. Modeling Rational Players I & II. *Economics and Philosophy* 3.2:179–214, 4.1:9–55.

———. 1989. Review of J. Harsanyi and R. Selten, "A General Theory of Equilibrium Selection in Games." *Journal of Economic Literature* 27.4:1171–73.

Binmore, Ken, and P. Dasgupta, eds. 1987. *The Economics of Bargaining*. Oxford: Basil Blackwell.

Born, Rainer, ed. 1987. *Artificial Intelligence: The Case Against*. New York: St. Martin's.

Bowden, B. V., ed. 1953. *Faster than Thought*. London: Pitman.

Burns, Eve. 1929. Statistics and Economic Forecasting. *Journal of the American Statistical Association* 24:152–63.

Cini, M. 1985. The Context of Discovery and the Context of Validation. *Rivista di Storia della Scienza* 2:99–122.

Collins, H. M. 1990. *Artificial Experts*. Cambridge, Mass.: MIT Press.

Craver, Earlene. 1986a. The Emigration of the Austrian Economists. *HOPE* 18.1 (Spring): 1–32.

———. 1986b. Patronage and the Direction of Research in Economics. *Minerva* 24.2–3 (Summer-Autumn): 205–22.

Dimand, R., and M. Dimand. 1992. The Early History of the Theory of Games. *HOPE*, this issue.
Dore, M., et al., eds. 1989. *John von Neumann and Modern Economics*. Oxford: Clarendon.
Dreyfus, Herbert. 1988. *Mind over Machine*, rev. ed. New York: Free Press.
Feyerabend, Paul. 1975. *Against Method*. London: NLB Books.
Fréchet, Maurice. 1953. Émile Borel, Initiator of the Theory of Psychological Games. *Econometrica* 21 (January): 95–127.
Gibbins, Peter. 1987. *Particles and Paradoxes*. Cambridge: Cambridge University Press.
Granger, C., and O. Morgenstern. 1970. *Predictability of Stock Market Prices*. Lexington, Mass.: D. C. Heath.
Harsanyi, John. 1967–68. Games with Incomplete Information Played by Bayesian Players. *Management Science* 14.3:159–82; 14.4:320–34; 14.5:486–502.
Harsanyi, J., and R. Selton. 1988. *A General Theory of Equilibrium Selection in Games*. Cambridge, Mass.: MIT Press.
Heims, Steve. 1980. *John von Neumann and Norbert Wiener*. Cambridge, Mass.: MIT Press.
Hodges, Andrew. 1983. *Alan Turing: The Enigma*. New York: Simon & Schuster.
Ingrao, Bruna, and Giorgio Israel. 1990. *The Invisible Hand*. Cambridge, Mass.: MIT Press.
Kline, Morris. 1980. *Mathematics: The Loss of Certainty*. New York: Oxford University Press.
Kreps, David. 1990. *A Course in Microeconomic Theory*. Princeton: Princeton University Press.
Kuhn, H., and A. Tucker. 1958. John von Neumann's Work in the Theory of Games and Mathematical Economics. *Bulletin of the American Mathematical Society* 64 (May): 100–122.
Leonard, Robert. 1992. Creating a Context for Game Theory. *HOPE*, this issue.
Lucas, William. 1967. A Counterexample in Game Theory. *Management Science* 13: 766–67.
———. 1969. The Proof that a Game May Not Have a Solution. *Transactions of the American Mathematical Society* 137.402 (March): 219–29.
Marget, Arthur. 1929. Morgenstern on the Methodology of Economics Forecasting. *Journal of Political Economy* 37 (June): 312–36.
Mehra, J., and H. Rechenberg. 1982. *The Historical Development of Quantum Theory*, vol. 3. New York: Springer Verlag.
Mirowski, Philip. 1986. Institutions as Solution Concepts in a Game Theoretic Context. In *The Reconstruction of Economic Theory*, edited by P. Mirowski. Boston: Kluwer.
———. 1989a. *More Heat than Light*. New York: Cambridge University Press.
———. 1989b. The Probabilistic Counter-Revolution. *Oxford Economic Papers* 41.1 (January): 217–35.
———. 1991. When Games Grow Deadly Serious: The Military Influence upon the Evolution of Game Theory. In *Economics and National Security*, edited by

Craufurd Goodwin. Durham, N.C.: Duke University Press.
———. 1992. Machine Dreams. University of Notre Dame, Economics Department Working Paper.
Morgenstern, Oskar. 1935. Vollkommene Voraussicht und Wirtschaftliches Gleichgewicht. *Zeitschrift für Nationalökonomie* 6.3 (August): 337–57.
———. 1937. *The Limits of Economics*. Translated by V. Smith. London: Hodge.
———. 1941. Professor Hicks on Value and Capital. *Journal of Political Economy* 49 (June): 361–93.
———. 1948. Demand Theory Reconsidered. *Quarterly Journal of Economics*. 62 (February): 165–201.
———. 1949. Economics and the Theory of Games. *Kyklos* 3.4:294–308.
———. 1962. On the Applications of Game Theory to Economics. *Giornale degli Economisti* 21.1–2:47–60.
———. 1976. The Collaboration between Oskar Morgenstern and John von Neumann on the Theory of Games. *Journal of Economic Literature* 14.3:805–16.
Morgenstern Papers, Oskar. Duke University Library. Durham, N.C. (OMP).
Morgenstern Papers: Diaries, Oskar. Duke University Library. Durham, N.C. (OMPD).
Nash, John. 1951. Non-Cooperative Games. *Annals of Mathematics* 54.2 (September): 286–95.
Pascal, Blaise. 1962. *Pensées*. Paris: Seuil.
Rellstab, Urs. 1992. New Insights into the Collaboration between John von Neumann and Oskar Morgenstern on the Theory of Games and Economic Behavior. *HOPE*, this issue.
Schelling, Thomas C. 1960. *The Strategy of Conflict*. Cambridge, Mass.: Harvard University Press.
Schotter, Andrew, ed. 1976. *Selected Economic Writings of Oskar Morgenstern*. New York: New York University Press.
Shankar, S. 1987a. The Decline and Fall of the Mechanist Metaphor. In Born 1987.
———. 1987b. *Wittgenstein and the Turning Point in the Philosophy of Mathematics*. Albany: SUNY Press.
———. 1988. *Gödel's Theorem in Focus*. London: Routledge.
Shubik, Martin. 1982. *Game Theory in the Social Sciences*. Cambridge, Mass.: MIT Press.
———. 1992. Game Theory at Princeton, 1949–1955. *HOPE*, this issue.
Steinhaus, H. 1960. Definitions for a Theory of Games of Pursuit. *Naval Research Logistics Quarterly* 7.2 (June): 105–8.
Tucker, A., and R. Luce, eds. 1959. *Contributions to the Theory of Games*, vol. 4. Princeton: Princeton University Press.
Turing, A. M. 1950. Computing Machines and Intelligence. *Mind* 59:433–60.
Ulam, Stanislaw. 1958. John von Neumann, 1903–1957. *Bulletin of the American Mathematical Society* 64.3 (May): 1–49.
———. 1976. *Adventures of a Mathematician*. New York: Scribners.

van Damme, Eric. 1987. *Stability and Perfection of Nash Equilibria*. Berlin: Springer Verlag.

von Neumann, John. 1958. *The Computer and the Brain*. New Haven: Yale University Press.

———. [1928] 1959. On the Theory of Games of Strategy. Translation in Tucker & Luce 1959.

———. 1961–63. *Collected Works*, 6 vols. Edited by A. H. Taub. New York: Pergamon.

von Neumann Papers, John. Library of Congress, Washington, D.C. (VNP).

von Neumann, J., and Oskar Morgenstern. 1964. *The Theory of Games and Economic Behavior*, 3d ed. New York: Wiley.

Wittgenstein, Ludwig. 1975. *Philosophical Remarks*, translated by R. Hargreaves and R. White. New York: Barnes & Noble.

———. 1978a. *Remarks on the Foundations of Mathematics*. Cambridge, Mass.: MIT Press.

———. 1978b. *Philosophical Grammar*. Berkeley: University of California Press.

Zermelo, Ernst. 1913. Über eine Anwendung der Mengenlehre auf die theorie des Schachspiels. *Proceedings, Fifth International Congress of Mathematicians* 2:501–4.

Part 2 The Diffusion of Game-Theoretic Ideas

Game Theory at Princeton, 1949–1955: A Personal Reminiscence

Martin Shubik

The report given here is clearly part of a Rashomon scenario. It presents a somewhat informal view of what was happening in game theory at Princeton during some of its early years seen through the eyes (and ego?) of a junior participant at the time. The view is impressionistic and undoubtedly biased in spite of my best intentions. The first draft of this essay was written quickly, from memory. This version has benefited from a reading of my diaries and from discussions with Lloyd Shapley and Herbert Scarf which enabled me to correct some outright errors and to more fully appreciate the Rashomon aspects of "eyewitness" reports.

I arrived in Princeton in the fall of 1949 with the express intention of studying game theory. I had sat in the library of the University of Toronto and attempted to read *The Theory of Games and Economic Behavior* in 1948 and was convinced that even though I scarcely understood the details of the mathematics, this was the right way to start to mathematize much of economics, political science, and sociology. The key feature was that it provided a language to describe precisely many of the key concepts of strategic analysis.

When I arrived in Princeton I found that my enthusiasm for the potentialities of the theory of games was not shared by the members of the economics department. Even the Princeton University Press, which as an academic publisher was meant to take reasonable risks with new scholarly enterprises, had required an outside subsidy of $4,000 before it would risk publication.[1]

There was Professor Morgenstern and his project with a few students, and there was the rest of the department. Although Morgenstern did give a graduate seminar in game theory, if one wished more than a single seminar it was necessary to become involved with the mathematicians

1. In defense of the press I must admit that the manuscript was extremely large.

and the activity at Fine Hall where a major research seminar was given and where there was a large group interested in game theory and its development.

The seminar given by Morgenstern that I attended had four students. They were Goran Nyblen, a brilliant young Swedish economist who eventually became paranoid and hung himself, but who did complete an interesting book on game theory and macroeconomics (Nyblen 1951). The second, William Young, went into industry immediately after graduate school, and unfortunately was dangerously alcoholic and died at an early age. The third, Djhangir Boushehri, decided to leave the graduate program before completing his degree and went on to a distinguished career as an applied economist (using more gamesmanship than game theory) at the International Monetary Fund. I was the fourth.

At Morgenstern's project were Maurice Peston, Tom Whitin, and Edward Zabel who were somewhat interested in game theory and operations research. Beyond that, game theory apparently had little impact on the economics department.

William Baumol raised questions about the value of the measurable utility assumption used in much game theory work at that time; outside of Princeton Karl Kaysen had questioned the worth of game theory in economics. The view was that in spite of the favorable reviews of Leonid Hurwicz and others this new mathematical bag of tricks was of little relevance to economics.

This view was put forward in particular by Jacob Viner whose favorite comment on the subject was that if game theory could not even solve the game of chess, how could it be of use in the study of economic life, which is considerably more complex than chess.

The graduate students and faculty in the mathematics department interested in game theory were both blissfully unaware of the attitude in the economics department, and even if they had known of it, they would not have cared. They were far too busy developing the subject and considering an avalanche of new and interesting problems.

Von Neumann was at the Institute for Advanced Study and Morgenstern was in the economics department. Then there was Albert Tucker in Mathematics who was actively interested. A host of junior faculty, visitors, and graduate students who in one form or the other were involved in some aspects of game theory, included Richard Bellman, Hugh Everett (recursive games), David Gale, John Isbell (absolute games), Sam Karlin, John Kemeny, Harold Kuhn, John Mayberry, John McCarthy, Harlan Mills, William Mills (four-person solution theory), Marvin Minsky, John Nash, Lloyd Shapley, Norman Shapiro, Laurie Snell, Gerald Thompson, and David Yarmish. Somewhat younger and arriving somewhat later were Ralph Gomory, Herbert Scarf, and William Lucas. John Milnor was an undergraduate and then a graduate student. Robert Aumann and many others came still later, after many of the earlier group had left.

The contrast of attitudes between the economics department and the mathematics department was stamped on my mind soon after arriving at Princeton. The former projected an atmosphere of dull business-as-usual conservatism of a middle league conventional Ph.D. factory; there were some stars but no sense of excitement or challenge. The latter was electric with ideas and the sheer joy of the hunt. Psychologically they dwelt on different planets. If a stray ten-year-old with bare feet, no tie, torn blue jeans, and an interesting theorem had walked into Fine Hall at tea time, someone would have listened. When von Neumann gave his seminar on his growth model, with a few exceptions, the serried ranks of Princeton Economics could scarce forbear to yawn.

I was hardly in a position to judge broadly at the time, but in retrospect, although some of us were primarily interested in game theory, the mathematics department as a whole had no special concern in the development of game theory *per se*. It was, to some extent, lumped with the budding developments in linear programming. Furthermore, topology, number theory, probability, differential equations, and many other domains of mathematics were actively being developed at Princeton at that time. The general attitude around Fine Hall was that no one really cared who you were or what part of mathematics you worked on as long as you could find some senior member of the faculty and make a case to him that it was interesting and that you did it well.

Although I did not appreciate it at the time, the book of von Neumann and Morgenstern could be regarded as four important separate pieces of work. They were (1) the theory of measurable utility; (2) the language and description of decision-making encompassing the extensive form and game tree with information sets, and then the reduction of the game tree to the strategic form of the game; (3) the theory of the two-person zero-sum game;[2] (4) the coalitional (or characteristic function) form of a game and the stable-set solution. Furthermore, with great care

2. More generally, constant-sum game.

von Neumann and Morgenstern had spelled out their attitude toward the relationship between game theory and economics and between dynamics and statics and the nature of what should constitute a solution.

Concerning economics they specified, "It will then become apparent that there is not only nothing artificial in establishing this relationship but on the contrary this theory of games of strategy is the proper instrument with which to develop a theory of economic behavior" (von Neumann and Morgenstern 1947, 2). Concerning the relationship between dynamics and statics, they observed,

> The next subject to be mentioned concerns the static or dynamic nature of the theory. We repeat most emphatically that our theory is thoroughly static. A dynamic theory would unquestionably be more complete and therefore preferable. But there is ample evidence from other branches of science that it is futile to try to build one as long as the static side is not thoroughly understood. On the other hand, the reader may object to some definitely dynamic arguments which were made during the course of our discussions. This applies particularly to all considerations concerning the interplay of various imputations under the influence of "domination." . . . We think that this is perfectly legitimate. A static theory deals with equilibria. The essential characteristic of an equilibrium is that it has no tendency to change, i.e. that it is not conducive to dynamic developments. An analysis of this feature is, of course, inconceivable without the use of certain rudimentary dynamic concepts. The important point is that they are rudimentary. In other words: For the real dynamics which investigates precise motions usually faraway from equilibria, a much deeper knowledge of these dynamic phenomena is required.
>
> A dynamic theory—when one is found—will probably describe the changes in terms of simpler concepts: of a single imputation—valid at the moment under consideration—or something similar. This indicates that the formal structure of this part of the theory—the relationship between statics and dynamics—may be generically different from that of classical physical theories. . . . Thus the conventional view of a solution as a uniquely defined number or aggregate of numbers was seen to be too narrow for our purposes, in spite of its success in other fields. The emphasis on mathematical methods seems to be shifted more towards combinatorics and set theory—and away from the algorithm of differential equations which dominate mathematical physics. (von Neumann and Morgenstern 1947, 44–45)

It is my belief that von Neumann was even more committed than Morgenstern to the idea of a solution as a set of imputations. He felt that it was premature to consider solutions which picked out a single point and he did not like noncooperative equilibrium solutions.[3] In a personal conversation with von Neumann (on the train from New York to Princeton in 1952),[4] I recall suggesting that I thought that Nash's noncooperative equilibrium solution theory might be of considerable value in applications to economics. He indicated that he did not particularly like the Nash solution and that a cooperative theory made more social sense. Professor Albert Tucker, in a personal conversation, informed me that in his conversations with von Neumann, von Neumann had displayed somewhat the same attitude to the single point solution, the value, proposed by Lloyd Shapley.

The von Neumann–Morgenstern stable-set solution is a sophisticated and sociologically oriented concept of stability. The authors (1947, 42) noted that they did not have a general proof of the existence problem and that if existence failed this would certainly call for a fundamental change in the theory. The attitude around Fine Hall was that if von Neumann conjectured that a stable-set solution always exists, the betting odds were that it did. Shapley and D. B. Gillies were looking for proofs or counterexamples, but although in the course of the next few years they were able to produce pathological stable sets (such as a solution in which you could append your signature as part of the stable set) an actual counterexample was not constructed until much later. William Lucas had worked on the von Neumann conjecture since Princeton and finally published his counterexample in 1968 (Lucas 1968).

Nash, Shapley, and I roomed close to each other at the Graduate College at Princeton and there was considerable interaction between us. In particular we all believed that a problem of importance was the characterization of the concept of threat in a two-person game and the incorporation of the use of threat in determining the influence of the employment of threat in a bargaining situation. We all worked on this problem, but Nash managed to formulate a model of the two-person bargain utilizing threat moves to start with. This was published in *Econometrica* (Nash 1953).

Prior to this work Nash had already done his important work on equilibrium points in n-person games in strategic form (Nash 1950). As I had

3. Even though the noncooperative equilibrium is frequently not unique.

4. I cannot verify the specific date.

read Cournot's work, I recognized that this was a great generalization of a concept that already existed in economics, the Cournot equilibrium point. Somewhat later, after Nash had completed the cooperative game model, he and I and John Mayberry collaborated in applying both the noncooperative and the cooperative models to duopoly with quantity strategies, such as Cournot models (Mayberry, Nash, and Shubik 1953). Later, with help from Shapley I extended the analysis to the Bertrand-Edgeworth models (Shubik 1955) and decided to do my thesis primarily utilizing the noncooperative solution applied to oligopoly problems.

There was considerable work going on on zero-sum game theory. My firsthand knowledge on this is less detailed and less accurate than on the other work because I felt that zero-sum games were not that interesting in application to economics.

Kuhn had studied two-handed poker and Nash and Shapley had considered three-handed poker. An expository article written by John MacDonald and John Tukey appeared in *Fortune Magazine* pointing out (among other things) that the concept of a bluff in poker was by no means merely psychological. They noted that even if one assumed totally passionless, bloodless individuals, they would bluff some percentage of the time in playing an optimal mixed strategy.

My own interests were directed primarily toward non-zero-sum games and both cooperative[5] and noncooperative theories. I will now move on to the cooperative theories.

The properties of stable sets including their intersection were originally considered by Gillies (1953) in his thesis. I believe that Shapley named the set of undominated imputations, the core of an n-person game. I was under the impression until I talked to Shapley that it was he who suggested considering it as a solution concept by itself. He pointed out to me that the idea of the core as a solution concept in its own right came up in our conversations when (as I was the only one in the group of us who was meant to know some economics), I observed that, in essence, the idea of the set of undominated imputations was already in

5. At Princeton I tried in vain to consider cooperative models that illustrated the role of money in an economy, feeling that it might have something to do with side payments and transferable utility. I made essentially no progress until about twenty years later when I worked on strategic market games rather than market games, the basic difference being that the former are process-oriented and the nature of money and financial instruments is part of the control system in an economic process. This is difficult to capture in the totally static and equilibrium-oriented analysis of cooperative games.

Edgeworth (1881) in his treatment of the contract curve, along with the idea of the replication of all players in order to study convergence.[6] In our conversations we originally were talking about stable-set solutions, but for the two-person Edgeworth bilateral monopoly the stable set and the core are identical. When we looked at the four-person game (two on each side) the distinction between the core and stable set eventually became clear. Sometime between 1952 and 1959 as we began to better understand what we were saying to each other and how the game theory compared with the work of Edgeworth, we understood the core as a separate solution concept.

As I was (and still am) mathematically weak, even though I recognized that the treatment in Edgeworth was of a game without transferable utility (currently referred to as NTU), as it was much easier to consider the game in side payment or transferable utility (TU) form, I first formulated it in that manner and eventually (Shubik 1959) was able to publish a simple proof of the convergence of the core of what I called "the Edgeworth game" to the competitive price.[7] Some years later, on a walk from Columbia University to downtown New York after Herbert Scarf had given a paper on an economy with a single dynamically unstable equilibrium point, I suggested to him that the core could be regarded as a combinatoric test for stability and I conjectured that the convergence of the core was probably true for NTU games. Scarf obtained a proof which he presented at a game theory conference in Princeton. It was published in the proceedings (which are difficult to locate). Somewhat later Debreu and Scarf (1963) obtained a somewhat more general proof of the existence and convergence of the core which was published in a readily available journal.

A fairly standard criticism of any attempt to interest the community of economists in cooperative game theory was that the representation of a game by a characteristic function entailed the implicit or explicit assumption of the existence of a magic substance or "utility pill" with a constant marginal utility to all traders. This assumption is called the TU assumption. The prevailing attitude of economists in the 1950s appeared to be that this assumption was so damaging as to make the application of cooperative game theory virtually useless.

6. The idea of replication is also in Cournot's treatment of duopoly being replaced by more competitors (1897).

7. I was helped by discussions with both Lloyd Shapley and, later, Howard Raiffa.

Von Neumann and Morgenstern had made this assumption not because it was a logical necessity but because it yielded a great simplification in the representation of an *n*-person game and enabled considerable calculation to be done which would have been far too complex in an NTU formulation. Shapley and Shubik (1953) pointed out that the assumption of TU was not only not needed, but that one could well define cooperative solutions to games where the preferences of individuals were represented only ordinally. A full development of this possibility did not take place until considerably later.

Shapley was concerned with developing a one-point solution for *n*-person games in coalitional form. He developed a set of simple but persuasive axioms which led to the selection of a value or a priori worth for each player. An interpretation of the Shapley value which eventually led me to suggest its application to the allocation of joint costs (Shubik 1952) was as a sociologically neutral expected combinatoric averaging over the marginal worth of an individual in all possible employments.

An immediate application of the value solution was to problems in voting, and Shapley and I (1954) collaborated in utilizing the value applied to a voting game to provide an index to measure the a priori voting power of an individual. At the time I had a few friends in the political science department who seemed to me to be more receptive to new ideas than members of the economics department.[8] William Ebenstein and Richard Snyder[9] encouraged us to consider sending this nonconventional approach to the *American Political Science Review* and much to my surprise it was accepted within a few months.

The Shapley value has been one of the most fruitful solution concepts in game theory. It generalizes the concept of marginal value and it, together with the Nash work on bargaining and the Harsanyi value, has done much in the last thirty years to illuminate the problems of power and fair division dealing with side payments and no side payments, fixed threats and variable threats, two individuals and many individuals.

Another informal activity at Fine Hall, although not immediately concerned with the mathematics of game theory, was of relevance. This was the many sessions (often at tea time) devoted to playing games (such as

8. I had the amusing experience of receiving from Friedrich Lutz a failing grade for a term paper in economic theory at the same time it was accepted for publication in *Econometrica* (Shubik 1952).

9. Richard Snyder invited me to edit a small book entitled *Readings in Game Theory and Political Behavior*. This, I believe, was the first booklet published explicitly on the theory of games as applied to political science.

go, chess, and kriegspiel) and to talking informally about paradoxical or pathological properties of games and the possibility of inventing games that illustrated these properties. Hausner, McCarthy, Nash, Shapley, and I (1964) invented an elementary game called "so long, sucker" where it is necessary to form coalitions to win, but this alone is not sufficient. One also has to double-cross one's partner at some point. In playing this game we found that it was fraught with psychological tensions. On one occasion Nash double-crossed McCarthy who was furious to the point that he used his few remaining moves to punish Nash. Nash objected and argued with McCarthy that he had no reason to be annoyed because a little easy calculation would have indicated to McCarthy that it would be in Nash's self-interest to double-cross him. We dubbed McCarthy's action as "McCarthy's revenge rule." If you are prevented from winning by a double-crosser, try to take the double-crosser with you.[10] "So long, sucker" still has not been fully analyzed, and the relationship between revenge and rational behavior still remains to be explored.

Some years later I formalized the rules for the dollar auction game (Shubik 1971) and considered its noncooperative and cooperative game solutions. The ideas for this illustration of a game with escalation or addiction in all likelihood may have had its origins (like the folk theorem concerning noncooperative equilibrium points in infinite horizon repeated games) in the informal sessions devoted to dreaming up paradoxical playable games.

To the best of my knowledge none of us at that time had formally considered experimental gaming in economics or political science, but the idea of experimentation was beginning to filter in from Mosteller and Nogee (1951) and later the book of Thrall, Coombs, and Davis (1954).

Unknown to me at the time was the breadth of the activity going on in linear programming. By 1947 von Neumann had conjectured the relationship between the linear programming problem and its dual and the solution of zero-sum two-person games. Gale, Kuhn, and Tucker started to investigate this more formally by 1948 and published their results in 1951. Kuhn and Tucker were also active in the development of nonlinear programming. The seminar at Fine Hall lumped the newly developing mathematics of game theory and programming together.

Although there was a beautiful link between the mathematics for the

10. Game theory still does not have an adequate formalization of revenge, resolve, bravery, morale, or any of the many other features that distinguish actual conflict from this type of abstraction.

solution of two-person zero-sum games, to a certain extent this link may have hindered rather than helped the spread of game theory understanding as a whole. For many years operations research texts had a perfunctory chapter on game theory observing the link to linear programming and treating linear programming and game theory as though they were one. The economics texts had nothing or next to nothing on the topic.

The extensive form of the game was of concern to several individuals at that time. Kuhn (1950) and Thompson (1953) were concerned with the representation of information and with the concept of strategy. I probably did not appreciate it sufficiently at the time, but the development of the notation for the extensive form had a considerable impact on decision theory and psychology. The ability to represent and analyze different information structures was a breakthrough of the first magnitude.

The role of von Neumann as a mathematician in the development of the theory of games is clear. The role of Morgenstern is less clear, and in my opinion underrated. As a former student of Morgenstern it can be argued that I am not in a position to give an unbiased estimate. Yet I feel that it is important to reconsider the nature of basic contributions. In many instances individuals discover new things and do not know the significance of what they have discovered. They see, but do not comprehend. They look, but have no vision.

One of the great virtues of Oskar Morgenstern was that he understood the significance of the theory of games. He was not a mathematician and on some occasions may not have even understood some of the work he espoused. But he was clearly aware of many of the big problems in economics and was energetic enough and visionary enough that he tried to do something about them even if he could not solve them himself. Thus, in particular, he recognized "perfect foresight" as a *bête noire,* and much of his concern for the development of the theory of games was to get rid of the paradox of perfect foresight.

There is little doubt that much of his "value added" came not merely from his own work, much of which was provocative and relevant (such as his book on the accuracy of economic measurement), but from his dedication to a talent hunt and to getting individuals to work on problems he thought were important. His influence on Wald, von Neumann, and his many mathematician friends was considerable. He wanted to encourage as much talent as he could to work on the problems he deemed important. In particular Morgenstern felt that it was important to try to guide first-class mathematical talent toward reconsidering the basic models and assumptions of economics, and his collaboration with von Neumann on game theory must be viewed in this light.

Although I have tried to give a view of part of a long and exciting campaign as remembered by one of the then very young campaigners, now forty years on, a few more comments must be made to set matters adequately in context. My time of observation at Princeton was from the fall of 1949 to spring of 1955. Even then the development was not merely at one location. RAND at that time was probably at least as important as Princeton, and many of the individuals named above worked at RAND or consulted there. In particular the work on stochastic games [11] and duels was of note. Although few of us at Princeton appreciated it, there was considerable activity at Michigan at the time as well.

Various individuals who were relevant to the development of the theory of games had already left Princeton before I arrived. There are some individuals at Princeton whose work I may have missed in this somewhat impressionistic and eclectic survey. No slight is intended; my main concern has been to try to indicate that, at least for some of us, this was a period of considerable excitement and challenge. New developments were taking place and somehow they seemed to be important even if we did not quite know why. We were present at the creation not only of game theory, but programming in general; we saw the development of the computer at the institute [12] as well as the development of other branches of mathematics.

When I consider the history of mathematical economics and the treatment of Cournot's great book, I am impressed by the growth in influence of the theory of games—not how little and how slowly, but how much and how fast.

The contrast between the Department of Economics and the Department of Mathematics at Princeton at that time has some lessons to teach. Besides Morgenstern there were some fine scholars in economics such as Viner and Baumol, but there was no challenge or apparent interest in the frontier of the science. Morgenstern was to some extent an inconvenience. To me, the striking thing at that time was not that the mathematics department welcomed game theory with open arms—but that it was open to new ideas and new talent from any source, and it

11. I was aware of Shapley's seminal paper (1953) and it was this that led me to formulate a non-zero-sum version called games of economic survival (Shubik and Thompson 1959).

12. With the help of the good offices of Alan Hoffman I managed to get time on the Johnniac to solve a 17 × 17 matrix game representation of a price duopoly model for my thesis.

could convey to all a sense of challenge and a belief that much new and worthwhile was happening.

References

Cournot, A. A. [1838] 1987. *Researches into the Mathematical Principles of the Theory of Wealth*. New York: Macmillan.

Debreu, G., and H. S. Scarf. 1963. A Limit Theorem on the Core of an Economy. *International Economic Review* 4 (October): 235–46.

Edgeworth, F. Y. 1881. *Mathematical Psychics*. London: Kegan Paul.

Gale, D., H. W. Kuhn, and A. W. Tucker. 1951. Linear Programming and the Theory of Games. In *Activity Analysis of Production and Allocation*, edited by T. C. Koopmans. New York: Wiley & Sons.

Gillies, D. B. 1953. *Some Theorems on N-Person Games*, Ph.D. diss. Department of Mathematics, Princeton University.

Kuhn, H. W. 1950. Extensive Games. *Proceedings of the National Academy of Sciences* 36.10 (October): 570–76.

Kuhn, H. W., and A. W. Tucker. 1953. *Contributions to the Theory of Games*, vol. 2. Princeton: Princeton University Press.

Lucas, W. F. 1968. A Game with No Solution. *Bulletin of the American Mathematical Society* 74.2 (March): 237–39.

Mayberry, J. F., J. Nash, and M. Shubik. 1953. A Comparison of Treatments of a Duopoly Situation. *Econometrica* 21 (January): 141–55.

Mosteller, F., and P. Nogee. 1951. An Experimental Measurement of Utility. *Journal of Political Economy* 59 (October): 371–404.

Nash, Jr., J. F. 1950. Equilibrium Points in N-Person Games. *Proceedings of the National Academy of Sciences* 36.1 (January): 48–49.

———. 1951. Noncooperative Games. *Annals of Mathematics* 54.2 (September): 289–95.

———. 1953. Two Person Cooperative Games. *Econometrica* 21 (January): 128–40.

Nash, Jr., J. F., and L. S. Shapley. 1950. A Simple Three-Person Poker Game. *Annals of Mathematics Study* 24:105–16.

Nyblen, G. 1951. *The Problem of Summation in Economic Science*. Lund: C. W. K. Gleerup.

Shapley, L. S. 1953a. Stochastic Games. *Proceedings of the National Academy of Sciences* 39.20 (October): 1095–1100.

———. 1953b. A Value for N-Person Games. In Kuhn and Tucker 1953.

Shapley, L. S., and M. Shubik. 1953. Solutions of N-Person Games with Ordinal Utilities (abstract). *Econometrica* 21.2 (April): 348–49.

———. 1954. A Method for Evaluating the Distribution of Power in a Committee System. *American Political Science Review* 48.3:787–92.

Shubik, M. 1952. A Business Cycle Model with Organized Labor Considered. *Econometrica* 20 (April): 284–94.

———. 1955. A Comparison of Treatments of a Duopoly Problem. *Econometrica* 23 (October): 417–31.

———. 1959. Edgeworth Market Games. In *Contributions to the Theory of Games*, vol. 4, edited by A. W. Tucker and R. D. Luce. Princeton: Princeton University Press.

———. 1971. The Dollar Auction Game. *Journal of Conflict Resolution* 15.1 (March): 109–11.

Shubik, M., and G. L. Thompson. 1959. Games of Economic Survival. *Naval Research Logistics Quarterly* 6.2 (June): 111–23.

Thompson, G. L. 1953. Signalling Strategies in N-Person Games, and Bridge Signalling. In Kuhn and Tucker 1953.

Thrall, Robert, C. H. Coombs, and R. L. Davis. 1954. *Decision Processes*. New York: Wiley.

von Neumann, J., and O. Morgenstern. [1944] 1947. *Theory of Games and Economic Behavior*, 2d ed. Princeton: Princeton University Press.

Game Theory at the University of Michigan, 1948–1952

Howard Raiffa

Not surprisingly, I was at Michigan in the mathematics department from 1948 to 1952—actually from the fall of 1946 through the spring of 1952—and participated in some of the collaborative research on the theory of games. It was an exciting time and I get vicarious pleasure thinking back in time to the sessions I enjoyed with my fellow student Gerald Thompson and my mentors Professors Arthur Copeland, Sr., and Robert Thrall. This historical account will be hopelessly biased because there is no paper trail of documents except those I have in my chaotic files of half-baked conjectures.

I was one of those returning G.I.s who descended on American campuses in the fall of 1946. I entered the University of Michigan as an undergraduate junior, having received some credits in the Air Force that were added to my one year of credits from CCNY where I started as a mathematics major. I grew up in the Bronx. Why Michigan? Well, I thought that I could support my wife and myself as an actuary and all one had to do was pass nine exams. Michigan was reputed to be the school that best prepared its students to take the first five of these exams.

I don't think that I heard the words "theory of games" in the entire academic year of 1946–47. But I did fall in love with the brain teasers posed in Whitworth's *Choice and Chance*, used in a course in probability theory. Although I passed the first three actuarial exams at the end of my first year of studies, I abandoned actuarial mathematics and decided to get my master's degree in statistics, which would draw more heavily on the theory of probability.

In the academic year 1947–48 once again there was no mention of game theory in the mathematics department at Michigan, at least as far as I was concerned. I took a course called "Foundations of Mathematics" with Professor Copeland, who taught in the R. L. Moore style: students are challenged to act like mathematicians, to convince themselves and others of the veracity of some plausible conjectures, to concoct starkly simple illuminating counterexamples, to generalize, to speculate, to abstract. No books were used. All the results were proved by the students. This is the mathematician's version of the case method of teaching. I became hooked. Even though I didn't know Leonard J. (Jimmy) Savage at the time, he also became enthralled in the same type of teaching program by being forced to act like a mathematician. I decided to become a pure mathematician and pursue a Ph.D. degree.

The ONR Project: Searching for Subs

So in the fall of 1948 I began my Ph.D. studies in pure mathematics. But one had to live, so I took a part-time job as a research assistant on a project sponsored by the Office of Naval Research (ONR), administered by the Engineering Research Institute. My supervisors were Robert Thrall (an algebraist), Cecil C. Craig (a statistician), and Arthur Copeland (a probabilist). One of my tasks was to attend conferences, mostly in Washington, D.C., in which applied mathematical problems were discussed by our clients in the Department of Defense and to formulate meaningful, tractable, mathematical problems. I remember working on problems of spatial search: patterns of search for elusive submarines. My associates on the project and I did not do two-party zero-sum equilibrium (or minimax) analysis, we just tried to find the best retort to a given strategy for the adversary. "Best" meant to maximize the probability of detection. It was not elegant stuff and I had my fill of elliptical integrals of the second and third kind. I learned that I hadn't given up actuarial math and statistics to become an applied classical mathematician, but conceptually I became intrigued with continuous two-person zero-sum theory before I knew about the minimax theory of von Neumann.

The Copeland Seminar and Solving Zero-Sum Games

In the fall of 1948 Professor Copeland started a baby seminar—by "baby" I mean that it was small enough that all of us interested folk could meet in his small office. We spent our time working through von Neumann and Morgenstern's account of the extensive-form game.

The theory was too abstract and nebulous for my tastes and Gerald Thompson and I got interested in grappling with some concrete two-

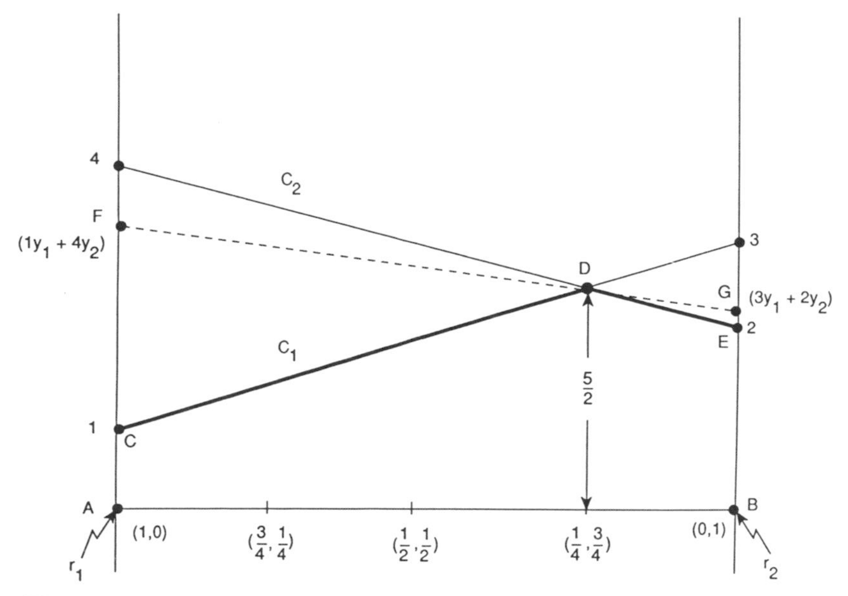

Figure 1

person zero-sum games with a finite number of pure strategies—small matrix games on the order of 5×5 or so. We did not know how to solve these games and this was our principal research task. We first concocted a graphical way to solve $2 \times n$ games. As an example, consider the 2×2 game with the following payoff matrix:

$$\begin{array}{cc} & \textit{Player 2} \\ & \begin{array}{cc} c_1 & c_2 \end{array} \\ \textit{Player 1} \begin{array}{c} r_1 \\ r_2 \end{array} & \begin{bmatrix} 1 & 4 \\ 3 & 2 \end{bmatrix} \end{array} \qquad \text{(Game 1)}$$

Player 1 has a choice of row r_1 or r_2 and player 2 has the choice of column c_1 or c_2. The payoff, for example, to player 1 corresponding to a pair of pure strategies (r_2, c_1) is 3. Player 1 is the maximizer; 2 is the minimizer. Figure 1 depicts the geometry. A point on the horizontal axis, such as (x_1, x_2) with $x_1 + x_2 = 1$ corresponds to the mixed strategy using weight x_i on r_i. The solid line (C, D, E) depicts the return player 1 can get if 2 uses her best retort against (x_1, x_2). As we see in the figure the strategy $(1/4, 3/4)$ maximizes the minimum function and guarantees to player 1 a return of $5/2$, the value of the game.

The same diagram can be used to illustrate geometrically the effects of player 2 using a mixed strategy y_1 on c_1 and y_2 on c_2. The payoff to player 1 corresponding to a mixed strategy (y_1, y_2) for 2 is shown by the dashed line going from F (at height $1y_1 + 4y_2$ against r_1) to G (at height $3y_1 + 2y_2$ against r_2). As y_1 moves from 1 to 0 the dashed payoff line rotates around the pivot point D from C_1 to C_2. The minimizing player 2 should choose y_1 such that the dashed payoff line is horizontal, yielding $5/2$ to player 1 regardless of what 1 does. So we see here the minimax theorem in its essence.

Gerald Thompson and I generalized this elementary idea, with Professor Thrall's help, to get an algorithm for solving finite zero-sum matrix games and shortly thereafter for solving linear programming problems. Figure 1 leaves out a significant complication. Suppose we add a third column to the game we have been discussing so it looks like this:

$$\begin{array}{c} \\ r_1 \\ r_2 \end{array} \begin{array}{c} \begin{array}{ccc} c_1 & c_2 & c_3 \end{array} \\ \begin{bmatrix} 1 & 4 & 0 \\ 3 & 2 & 5 \end{bmatrix} \end{array} \qquad \text{(Game 2)}$$

Figure 2 illustrates the ensuing discussion. Observe that C_3 is below C with coordinates $(1, 0; 1)$ and above D with coordinates $(1/4, 3/4; 5/2)$. Hence the minimum function must now add the piercing point H, delete the point C, and add the point A so that the new minimum function would be characterized by $\{A, H, D, E\}$.

But C_3 is also below C and above E, so why not a piercing point on a line connecting C and E? The answer, of course, is that points C and E are not adjacent on the minimum function. When we generalize to more than two rows and we don't have a simple geometric picture in front of us, we must not only keep a current list of the points that characterize the minimum function, as we build it up by introducing columns sequentially, but we must keep in mind which points are adjacent. This is not difficult to do, however, if we keep track not only of the coordinates of the extreme points of the minimum function but the lines (or hyperplanes in higher space) that define those points. Thus, for example, point D in figure 2 would be characterized by a *double description:* $(1/4, 3/4; 5/2)$ and cutting planes $\{C_1, C_2\}$. By this means it is possible even in higher spaces to ascertain whether points are adjacent. This work was informally published in our ONR project report dated September 1950 (see Raiffa, Thompson, and Thrall 1950). When this report was written we

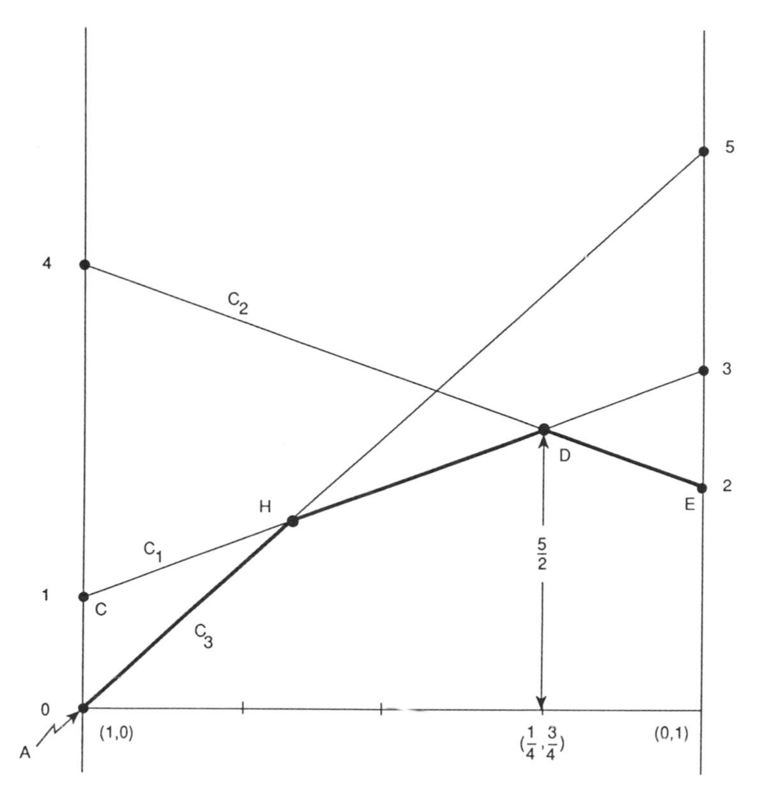

Figure 2

made reference to work at RAND in 1948 that could be used to solve games but we were not aware of the contemporaneous work of Dantzig on the simplex method.

In the fall of 1950 we learned that the minimax analysis of the two-person zero-sum game was a special case of a linear programming problem and that our algorithm for solving games could be trivially extended for solving linear programming problems as well. We also learned at a national meeting, where we reported our results, that T. S. Motzkin had explored a similar technique for solving linear inequalities and we agreed to jointly publish our work under the title of "Double Description Method" (see Motzkin, Raiffa, Thompson, and Thrall 1953).

Signaling Strategies

In the academic year 1950–51 Gerald Thompson and I worked on different problems: he on the extensive-form game and I on the two-person non-zero-sum game. The Copeland seminar in 1949–50 paid great attention to the extensive and normal forms of a game, and we learned about Harold Kuhn's work on the strategic equivalence of mixed and behavioral strategies for games of perfect recall. Thompson became interested in the simplified versions of bridge that can be thought of as a north-south player against an east-west player. Each player, however, has a split role. The north-south player at one information set may not remember what choice that composite player made at an earlier information set. It is not a game of perfect recall, and the north-south player, for example, may want to correlate his/her randomized choices at different information sets. Kuhn showed that in games of perfect recall there is no loss of strategic control if a player restricts him- or herself to behavioral strategies. Essentially players can simply randomize independently at each information set without having to correlate their randomized choice. Not so in games without perfect recall. Thompson focused his attention on those information sets—the so-called signaling ones—which prevented perfect recall. He generalized the notion of behavioral strategies to allow for signaling and showed that no strategic control is lost by restricting the class of mixed strategies to a class of strategies a bit more complicated than behavioral strategies: a perfectly intuitive result once one keeps in mind the game of bridge (see Thompson 1953).

The Non-Zero-Sum Game

Von Neumann and Morgenstern in their epic classic cleaned up all conceptual problems for the zero-sum game with finite numbers of pure strategies. For each game there was a value v of the game; the row player could guarantee himself at least v by playing a maximin strategy; and the column player could guarantee that the row player would get no more than v by playing a minimax strategy. Any paired selection of a maximin and minimax strategy was in equilibrium and any equilibrium pair yielded a maximin and minimax strategy. How neat. But the non-zero-sum domain defied orderly analysis.

Late in the academic year 1949–50 I worked on the non-zero-sum game for our ONR project. I did what a lot of researchers did at the time, quite independently of each other. I investigated all qualitatively different 2×2 bi-matrix games and tried to identify interesting subspecies. Many types of anomalies appeared, stemming mostly from the existence of multiple equilibria pairs yielding different payoffs. I thought we could get consensus about what we might mean by a game having a "solution" for the case in which there was a unique equilibrium pair that jointly dominated all other equilibria pairs. For example, consider the game depicted below:

$$\begin{array}{c} \\ r_1 \\ r_2 \end{array} \begin{array}{c} \begin{array}{cc} c_1 & \quad c_2 \end{array} \\ \begin{bmatrix} (0, -1000) & (10, 8) \\ (12, 10) & (0, 0) \end{bmatrix} \end{array} \qquad \text{(Game 3)}$$

The payoff in cell (r_2, c_1), for example, yields 12 and 10 to players 1 and 2 respectively. The pair (r_2, c_1) is in equilibrium; so is (r_1, c_2). But (r_2, c_1) jointly (Pareto) dominates (r_1, c_2) and therefore I argued that (r_2, c_1) should be the solution of this game. But I wasn't happy. Solution in what sense?

Player 2 could think to herself, "I'm in deep trouble if I take c_1 and he takes r_1. But he's no dope; he might think I'd be afraid to take c_1 and therefore he might be inclined to take r_1. But now that I can give some rationalization for him to take r_1, I certainly should stay away from c_1." The argument becomes cyclic. The stability of the equilibrium (r_2, c_1) gets destroyed once one gives some probability of a deviation from that equilibrium. Wise advice, jointly given openly to both players, would suggest r_2 for 1 and c_1 for 2. But is c_1 wise advice to player 2 in isolation?

I started to worry about the question "solution in what sense?" Certainly, game theory was not meant to be a descriptive theory or predictive theory of what the vast majority of untutored players do. How about predictive power among the tutored? Among students previously exposed to game-theoretic ideas and knowledgeable about equilibrium analysis, it turned out empirically that c_1 was still a poor choice for column players. I began to feel uncomfortable about the efficacy of the equilibria theory as either a descriptive or prescriptive theory. Something else had to be added.

In my investigation of different qualitative types of 2×2 non-zero-sum matrix games I stumbled over the game that is now commonly called the prisoner's dilemma. In that game, as I characterized it, each player has a single dominant strategy but the resulting pair of dominating strategies is not jointly (Paretian) efficient. To be concrete, consider the game

$$\begin{array}{c} \\ r_1 \\ r_2 \end{array} \begin{array}{cc} c_1 & c_2 \\ \left[\begin{array}{cc} (5,5) & (-10,10) \\ (10,-10) & (0,0) \end{array}\right] \end{array} \qquad \text{(Game 4)}$$

Row r_2 is best against c_1 or c_2 (and hence dominant) and column c_2 is best against r_1 or r_2 (and hence dominant), yet (r_2, c_2) is jointly inferior to (r_1, c_1). The smart thing, according to theory, is to take (r_2, c_2). Two stupid players, however, do better than two smart players. Still, I wrote in 1950 that (r_2, c_2) "solved" this game; I was comfortable with that analysis from a descriptive and prescriptive perspective. The trouble came when that game was played a finite number of times, say, twenty times. In the extensive form of the game it was clear that on the twentieth trial each player should defect (i.e., take the second row or column); so working backwards, each player should defect on trial 19; and so on. The only equilibrium strategy is for the players to defect immediately on trial 1 and keep going for twenty trials. I felt dismayed by this result. Descriptively this did not happen with either untutored or tutored subjects, as I found out in my informal experiments in 1950. All this was documented in progress reports I prepared for our ONR project and summarized in project report R-30, dated June 1951 (see Raiffa 1951). In that report I asked, "What prescriptive advice would I give to one player playing against another tutored player?" I was grappling with an approach that I later dubbed the "asymmetric prescriptive/descriptive approach"—i.e., giving *prescriptive* advice to one player on the basis of *descriptive* modeling of the other players.

In 1950 when this work was being done I was aware of Wald's (1950) work on statistical decision theory. Wald was not a subjectivist, but in a purely formal way he examined what he called "Bayesian Decision Rules" as a means of finding undominated (Paretian) decision rules for some common statistical decision problems. In desperation, I resorted, in the 1951 report, to act as a subjectively oriented decision analyst: to assign a subjective probability distribution over a reasonable class of strategies for the other player and to optimize my own (or my client's) retort against that distribution. At the time I was not aware of the Chicago School (L. J. Savage, H. Chernoff, H. Rubin) who were taking the subjective viewpoint seriously in statistical decision theory. In classical game theory, however, even today there is no role given to the use of subjective probability in formal analysis.

I chose not to seek wider publication of these results because I viewed them as negative. I was disillusioned with the hope of gaining significant insights into large classes of non-zero-sum games based on variations of equilibria theory.

In *Games and Decisions* by Luce and me (1957) I wrote the first draft of chapter 5 on the two-person non-zero-sum noncooperative game and based it on the first part of my 1951 ONR report. In that ONR report I analyzed game 4 above and reported that that game "has the same structure as a game first suggested by Dr. Dresher at the RAND Corporation." In my bibliography, however, there was no reference to any paper by Dresher. I learned about Dresher's work from Thrall only after developing my own analysis of that game. As far as I knew at the time, the existence of this perverse game was so close to the surface that it was folk knowledge. In Luce and Raiffa 1957, we give credit to A. W. Tucker for the prisoner's dilemma *interpretation* of game 4 and cite a report by M. M. Flood (1952) who wrote a delightful account of the game repeated one hundred times. Flood recorded the articulated thought processes of two players in that frustrating game. In 1953, Flood gave a seminar at Columbia University (where Duncan Luce and I both were at the time) on the iterative prisoner's dilemma game, and I congratulated him on a beautiful presentation. After *Games and Decisions* was published, Flood wrote me a letter asserting that I got the history wrong. He claimed that I first learned about the game in 1953 at his seminar and he should have gotten the priority claim instead of Tucker. He was demonstrably wrong in thinking that I learned about the game in 1953, but he, along with Dresher (both were at RAND), might very well have thought about the game at the earliest documentable date. All I know is that economists probably knew for a long time before 1950 that jointly dominating strategies could be Paretian inefficient. In my 1950 notes I referred to the plight of wheat farmers, each of whom is better off with full production even though this would depress prices to the detriment of all.

Two-Person Cooperative Games

In the spring of 1950 I felt frustrated. My work on the two-person non-zero-sum noncooperative game was unyielding. There were too many inadequacies in a theory based on just equilibria analysis. As far as I was concerned the theorizing I did lacked descriptive or prescriptive applicability. I distinctly remember attending a seminar at Michigan given by Professor Haber on arbitration of labor-management problems and getting all excited about a new idea. Imagine that two of my game-theoretic friends had to play an abstract two-person non-zero-sum game; suppose that, in fear that they would fare poorly, they approached me to arbitrate a solution: to propose a binding resolution of the game. What principles of efficiency and fairness would I want to impose? At that time—before it was published the next year—I knew about Kenneth Arrow's *Social Choice and Individual Values*, and I discussed his work at a seminar jointly sponsored by Clyde Coombs in social psychology and Robert Thrall in mathematics. I became enamored of the use of the axiomatic method in sorting out the desiderata for choice mechanisms. In the summer of 1950 I also learned about the Bargaining Problem of Nash but I objected to his notion of *solving* the game. I thought his "solution" did not have descriptive or prescriptive credence. But if it were interpreted as an *arbitration scheme*, then it made sense to me. Only I didn't like his axiom of the Independence of Irrelevant Alternatives in the context of arbitration. My research on arbitrated resolutions of two-person non-zero-sum games is given in my ONR summary report of 1951, published in a 1953 paper and rewritten for *Games and Decisions* in 1957 as chapter 6. I reported on this research in the fall of 1950 at Professor Tucker's seminar at Princeton. There was a suggestion made at that seminar that Nash and I collaborate, but Nash was not easy to work with, to put it mildly.

In the early spring of 1951 after I drafted the ONR report summarizing my research for the project, I settled down to write a thesis on linear-normed spaces, when my mentors, Professors Copeland and Thrall, suggested out of the blue that I submit my project report as my Ph.D. thesis. Was I surprised! So I was finished before I started, and this led me down the path of applied rather than pure mathematics. A happy choice for me in retrospect.

Postscript

Over the years I have drifted away from classical game theory. The exclusive emphasis on joint rationality and common knowledge and the dependence on equilibria analysis seems to me to limit the applicability of the field both descriptively and prescriptively. I think of myself as a professional decision analyst: someone concerned with helping myself, or my client, make wise choices in uncertain environments and in the presence of conflicting objectives. In interactive, gamelike contexts, some uncertainties stem from the actions of others who may be thinking what you are thinking. But this interactive, reflexive thinking process rarely goes deeper than one level—except perhaps in highly simplified textbook games—and game-theoretic equilibrium is a poor predictor of the actions of others. Given this observation, the approach I favor is to assess subjective probability distributions for the actions of others based on behavioral models and then to optimize against this distribution. This approach is not in the spirit of classical game theory but the domain of applicability is far wider because it does not depend on cyclical ultrarationality and the binding constraints of common knowledge. My uncomfortable feeling about equilibria analysis as a descriptive or prescriptive tool—but not as a normative theory—has not changed from my feelings of forty years ago at the University of Michigan.

References

Flood, M. M. 1952. Some Experimental Games: Research Memorandum RM-789. Santa Monica: RAND Corporation.

Kuhn, H., and A. W. Tucker, eds. 1950. *Contributions to the Theory of Games*, vol. 1. Princeton: Princeton University Press.

———. 1953. *Contributions to the Theory of Games*, vol. 2. Annals of Mathematical Studies 24. Princeton: Princeton University Press.

Luce, Duncan, and Howard Raiffa. 1957. *Games and Decisions*. New York: Wiley.

Motzkin, T. S., Howard Raiffa, G. L. Thompson, and R. M. Thrall. 1953. The Double Description Method. In Kuhn and Tucker 1953.

Nash, J. F. 1950. The Bargaining Problem. *Econometrica* 18 (April): 155–62.

Raiffa, H. 1951. Arbitration Schemes for Generalized Two-Person Games. Engineering Research Institute, University of Michigan, Report No. M720-1 R-30. June.

———. 1953. Arbitration Schemes for Generalized Two-Person Games. In Kuhn and Tucker 1953.

Raiffa, H., G. L. Thompson, and R. M. Thrall. 1950. An Algorithm for the Determination of All Solutions of a Two-Person Zero-Sum Game with a Finite Number of Strategies. Engineering Research Institute, University of Michigan. Report No. M720-1 R-28. September.

Savage, L. J. 1951. The Theory of Statistical Decisions. *Journal of the American Statistical Association* 46.253 (March): 55–67.

———. 1954. *The Foundation of Statistics*. New York: Wiley & Sons.

Thompson, G. L. 1953. Signalling Strategies in N-Person Games, and Bridge Signalling. In Kuhn and Tucker 1953.

Wald, A. 1950. *Statistical Decision Functions*. New York: Wiley & Sons.

Mathematizing Social Science in the 1950s: The Early Development and Diffusion of Game Theory

Angela M. O'Rand

> The man of letters is increasingly finding, to his dismay, that the study of mankind proper is passing from his hands to those of technicians and specialists. The aesthetic effect is admittedly bad: we have given up the belletristic "essay on man" for the barbarisms of a technical vocabulary, or at best the forbidding elegance of mathematical syntax. What have we gained in exchange?—Abraham Kaplan, "Sociology Learns the Language of Mathematics"

The publication of *The Theory of Games and Economic Behavior* in 1944 was among several related developments in the social sciences in the 1940s that signaled an acceleration in the spread of mathematical thinking. Although the implementation of such approaches had been encouraged for several decades—indeed the Social Science Research Council was founded in the 1920s under the leadership of Charles Merriam with precisely this purpose—progress in this direction had been slow and uneven across the social sciences. Economics was the most advanced in this respect because of the greater correspondence (i.e., isomorphism) between formal economic theory and mathematical representations. On the other hand, sociology and political science had developed verbally rich but more conceptually ambiguous theory not as readily translated into mathematics (Simon 1957, Lazarsfeld 1954, Coleman 1964). Nevertheless, the confluence of historical and intellectual developments at this time led to a "watershed" in the social sciences (Coleman 1986, 1314)—that is, to a historical transition period constituted of multiple fronts for the development of mathematical thinking.

This article examines critical features of the social and cognitive landscape of the social sciences in the period between approximately 1945 and the early 1960s during which the von Neumann–Morgenstern research program in game theory is presented as one of several "research fronts" developing with relative success across the social sciences. A research front is defined as a highly productive, relatively exclusive, casually organized, and usually short-lived problem-solving area in science characterized by rapid development and the pursuit of originality (O'Rand 1991).[1] It may be conceived as the immediate market within which "incessant attempts at getting attention and imposing ideas and concepts on colleagues" are traded (Whitley 1984, 26). As such, this period of game theory is viewed both as a cognitive agenda, i.e., as a unique research program in the rational sense (Lakatos 1978), and as a socially distinct research school, i.e., as a relatively localized configuration of individuals, institutions, and resources following models of research schools in other sciences that have been studied by historians (see Geison 1981).

Together the rational and social characteristics of the game theory program constituted a technical or *epistemic culture* (Knorr Cetina 1991), a technical system of scientific thought style, social relationships, and localized communication and collaboration patterns that distinguished it from other developing programs in the social sciences. The mathematical (combinatorial) solution—the proof—was central to the rational system of game theory which developed in a *transepistemic* network of institutions and researchers moving among this circuit of institutions. I borrow the concept of transepistemic arenas of research from Knorr Cetina (1982) to denote that the research enterprise necessarily involved both scientists and nonscientists and encompassed "arguments and concerns of a 'technical' as well as a 'non-technical' nature" (Knorr Cetina 1982, 101). Indeed, the research arena for game theory comprised a network of universities, institutes, private and public funding agencies, nonprofit corporations, and conference/workshop sites that served both

1. Research fronts are the "growing tips" of the sciences and thus distinguished from the broader disciplines, research traditions (Laudan 1977), or intellectual fields (Whitley 1984). The latter constitute the continuity of thought, style, and general methodology that distinguish relatively large-scale and long-lived socioconceptual heritages handed down over generations of scientists. These, by contrast to the research front, are highly organized intellectual and social organizations with predominant formal features exerting dispersed control over the scientific process. In economics, the general theory of rational behavior can appropriately be designated a research tradition (see Suppes 1961; Harsanyi 1977). Game theory, one branch of this tradition, can be designated a "research front" over the initial period of its active development both cognitively and socially during the post–World War II period. It was highly productive but relatively exclusive and geographically localized across a set of institutions (e.g., Princeton, RAND, Office of Naval Research, Stanford's Center for Advanced Study in the Behavioral Sciences, etc.) and in some competition with other programs for resources and influence.

to demarcate it as a cognitive-social region of the social sciences and to bring it into episodic contact with alternative research fronts.

Chief among these alternative programs for the mathematization of the social sciences were Kurt Lewin's field theory centered at MIT and the media sociology centered at Columbia University's Bureau of Applied Social Research under the direction of Paul F. Lazarsfeld. The technical core of each of these programs was not the mathematical proof, as in the case of game theory, but the social experiment and the social survey, respectively. And while all three programs were connected in a circuit of epistemic centers (O'Rand 1991)—centers of intellectual and social gravity—along which many of the same social scientists traveled, the extent of mutual cognitive influence among them over this period was remarkably small.

This socio-cognitive landscape emerges with the analysis of selected bibliographic references from the social sciences between 1945 and 1965,[2] biographical and autobiographical accounts drawn from pub-

2. The narrative that follows is developed from an analysis of approximately 380 articles, anthologies, monographs, and miscellaneous reports and approximately 500 social science citations. An invaluable baseline source for the reconstruction of the group dynamics front was Strodtbeck and Hare's (1954) bibliography of small-group research. This is a relatively exhaustive collection of bibliographic references (N = 1407) in this broad area spanning the period between 1900 and 1953, 62 percent of which were published after 1944. The compilers engaged nine judges to check those contributions they considered important and utilized the extent of agreement among judges on all items to determine the most important contributions. This facilitated the task of assessing core research areas and alternative "research fronts" coexisting with the game theorists over the early period under study.

In addition, two dozen serials were examined over the period of observation to categorize influence patterns observable in citations, resource support footnotes, and institutional locations. The serials included in this study were *American Anthropologist; American Economic Review; American Journal of Sociology; American Sociological Review; American Political Science Review; American Psychologist; Annals of Mathematical Studies; Behavioral Science; Econometrica; Human Relations; Journal of Abnormal and Social Psychology; Journal of Experimental Psychology; Journal of Political Economy; Journal of Social Psychology; Journal of Social Issues; Journal of Conflict Resolution; Management Science; Operations Research; Proceedings of the National Academy of Science; Psychological Bulletin; RAND Publications; Social Forces; Sociometry.*

The *Social Science Citation Index* for the period 1956–65 (Institute for Scientific Information 1989) was utilized to estimate relative citation rates to evaluate the Strodtbeck and Hare bibliography and to locate the "core" game theory literature disseminating beyond the research front. The most highly cited game theory documents over this period were von Neumann and Morgenstern 1944, 1947, 1953, 1957; Luce and Raiffa 1957; Shapley and Shubik 1954; and McKinsey 1952, in order of rating. However, if each of the four editions of *The Theory of Games and Economic Behavior* is treated separately, the Luce and Raiffa book is the most highly cited (with 169 citations).

lished sources and interviews,[3] and the Oskar Morgenstern papers.[4] The fundamental argument of this analysis is that the course of development of game theory during this period is better understood when elements of the larger historical and intellectual context are taken into account. Development is conceived at the cognitive level as the coherent extension and "hardening" of programmatic theory (Lakatos's notion of a progressive research program) and at the social level as diffusion, or the incorporation of the program by neighboring social scientific programs, demonstrating the program's broader epistemic credibility (Fuller 1989; Latour 1987).

Intellectual Migrations: Mathematizing the Social Sciences

Kaplan's somewhat bittersweet appraisal of the state of the social sciences in the early 1950s (see epigraph) introduces his review of alternative research programs that had developed in the recent period to incorporate mathematical thinking in the social sciences. Chief among his examples were Clark Hull and the Yale Institute of Human Relations' approach to rote learning invoking geometric principles, Stuart Dodd's "ambitious" but "misguided" quantitative systematics for the social sciences, Rashevsky's social behavioral models of phenomena such as imitation and status distribution derived from mathematical biology, Zipf's probability model of the "principle of least effort" in linguistics, Thurstone and Spearman's and, later, Guttman's measurement scaling methods, Lewin's topological field theory in the social sciences, Wiener's cybernetics, Lazarsfeld's latent structure analysis, and von Neumann and Morgenstern's game theory (Kaplan 1956). He did not consider these alternative programs to be necessarily in direct competition over

3. Biographical sources, including the *Dictionary of Scientific Biography* and *American Men and Women of Science*, were used to examine career patterns and geographical/institutional mobility patterns. In addition, several autobiographical and interview collections were particularly invaluable: Jacques Cattell 1975; Coleman 1986, 1990a, 1990b; Dieubonne 1976; Fleming 1969; Lazarsfeld 1969; Lindzey 1989; Mandler and Mandler 1969; McGill 1988; Ulam et al. 1969; Swedberg 1990; and Berger 1990.

Finally, interviews and discussions with Kurt Back, Max Woodbury, Theodore Caplow, Ed Borgatta, William Riker, and Phil Costanza were especially illuminating, as were comments by conference participants on 5–6 October 1990, who also contributed to this volume.

4. Oskar Morgenstern Papers (OMP), Boxes 41, 42, and 47, Special Collections Department, Duke University Library, Durham, North Carolina.

the analysis of the same social phenomena. Rather he viewed them as different languages fitted to distinctive problem definitions which would collectively contribute to the emergence of systematic general theory.

This pluralist position reiterated Herbert A. Simon's second canon of strategy for the construction of social science models—that a plurality of mathematical models was needed (Simon 1954). Simon, in particular, sensed that centripetal forces were inducing a process of integration among the social sciences. He believed that "large events" (wars, elections, depressions) fostered such convergences with implications for scientific change and that opportunity knocked at that moment for the successful introduction of mathematics to social and political theory. Whether or not an actual convergence was occurring is more readily judged from our perspective four decades later. Yet, his intuitive sense at the time, seemingly widely shared, regarding the influences of intellectual migration on theory change and disciplinary development is consonant with earlier historical episodes documented in the sciences (Urban 1982). The emergence of psychology as a discipline in the nineteenth century (Ben-David and Collins 1966), the shift from optical to radio astronomy in England (Edge and Mulkay 1976), the formation of the phage group by Max Delbruck and its impact on the emergence of molecular biology (Fleming 1969; Mullins 1972), and the origins of modern cell biology in the United States (O'Rand 1986) have all been associated with events such as wars or intellectual migrations by scientists between nations and disciplines.

The large events of the mid-twentieth century produced, among other things, the geographical migration of European intellectuals to the United States (e.g., Fleming and Bailyn 1969) and the institutionalization of "big science" (Price 1963, Abir-Am 1982), even a relatively "big social science" after the war. Big social science consisted of multi-institutional networks of social scientists and their respective federal and private sponsors committed to the solution of shared problems. Game theory's contribution to the science of the strategic game lent itself to postwar concerns over operations research, management science, and world politics. And von Neumann's seminal contribution to its foundation and subsequent arrival in the United States to work with Morgenstern established a transepistemic arena of activity linking mathematicians, social scientists, strategic planners, and increasingly, wider social science communities over time (Ulam et al. 1969).

Emergence and Development of American Game Theory

Von Neumann's meeting with Oskar Morgenstern in 1939 was referred to by the latter as a "gift from heaven" (Weintraub 1985, 21). Their mutual interests in strategic behavior joined von Neumann's mathematical prowess with Morgenstern's interest in transcending strictly individual-level analyses in the building of a new economics. They united in an intellectual contract resulting in their 1944 publication and in an institution-building endeavor to construct a full-blown discipline (Aumann 1987, Mirowski 1991). They developed one of the first social science–centered mathematical innovations (Luce and Raiffa 1957).

The von Neumann–Morgenstern program developed into a research front, following the definition provided earlier, by 1950. Its status and development as a research front so defined and its socio-cognitive location in the transepistemic landscape of the social sciences during the post–World War II period is the subject of this analysis. The research program introduced in *The Theory of Games and Economic Behavior* and extended over the next two decades was an ambitious extension of rational behavior theory to account for a wider array of socioeconomic phenomena related to competition and cooperation. The idea of a scientific research program is drawn from Lakatos (1978) who proposed, in opposition to Kuhn's view of theory change in the sciences, that the appropriate unit of scientific development or change is not the grand paradigm but the scientific research program, or the developing cluster of theories attached to a problem domain. The program's structure consists of a tenacious hard core of untested assumptions and heuristics, or a set of rules (usually in the form of instruments or mathematical apparatus), prescribing the theory-building process deployed in a belt of auxiliary hypotheses organized to protect the "hard core." The hard core of general equilibrium analysis has been described by Weintraub (1985).

The specific aim of game theory was to find the mathematically complete principles that define choice optimization ("rational behavior") for actors in a social economy and to derive from these principles the general characteristics of that choice process. The cornerstone (starting point) of the theory was the assertion and proof of the minimax theorem of the two-person zero-sum game completed by von Neumann sixteen years earlier. The theorem asserted optimal strategies for two players and the unique numerical value (the minimax value) of the game. The von Neumann–Morgenstern theory proposed the solution for cooperative games, where all possible partitions of players into two coalitions were defined. It also established the imputations (stable set) derived from the endogenous characteristics (organization) of the players.

The method employed by the authors in applying the theory consisted of taking an economic problem (they chose voting and issues related to monopoly or oligopoly in market contexts), formulating it as a game, finding the solution, and then translating the game back into economic terms (Aumann 1987, 466). This approach contributed to its ultimate appeal, particularly in economics and political science.

Both von Neumann and Morgenstern were convinced that mathematics rather than physics or mechanics provided the proper toolbox for research in the "complicated" social sciences. They argued that games of strategy were the proper analogues for economic situations and that a mathematically grounded game theory provided the solution. Accordingly, the mathematical solution was viewed as the cornerstone of theory-building based on the argument that "combination is at the very bottom of the foundations of mathematics—when something is settled combinatorically, it is really decided."[5] Thus, the primary intellectual currency in this program was the proof usually exchanged as thought experiment in paper-and-pencil form or game-board design.[6]

The decade and a half that followed the publication of *The Theory of Games and Economic Behavior* consisted of a high level of informal and formal interaction among a tight circle composed largely of mathematicians and economists working on clarifications and solutions for cooperative and noncooperative conditions of strategic games. Considerable informal interaction occurred at strategic sites (principally at Princeton) on a circuit of meeting places. Yet, the publication patterns were largely of the single- or dual-author type. Program solutions that developed from these interactions and collaborations over the decade or so after 1944 pertained to problems related to equilibrium, incomplete information, repeated games, the prisoner's dilemma, value (power) indexes, among others. These developments appear in a limited set of serials (including the Princeton *Annals of Mathematics Studies*, *Econo-*

5. Notes dated 2 January 1947, Box 41, subject file Game Theory: Further Papers, 1948–1965, OMP.

6. John Nash to Morgenstern, unpublished paper on the game "hex" dated May 1951, and Lloyd Shapley and Martin Shubik to Morgenstern, unpublished paper on "Solutions of N-Person Game with Ordinal Utility" dated September 1952, Box 42, OMP.

metrica, *Proceedings of the National Academy of Sciences*, and report collections attached to the Office of Naval Research and the RAND Corporation). However, beyond the von Neumann–Morgenstern book, the pieces of this literature that moved into the wider social-scientific domain before the end of the 1950s were few and far between: chief among them were McKinsey 1952, Shapley and Shubik 1954, and Luce and Raiffa 1957. Accordingly, the program-building literature itself did not serve as a fundamental disseminator of the program.[7]

Aumann (1987), Max Woodbury (1990), and biographers of von Neumann and Luce confirm that the literature was not the primary arena for theory diffusion. Two reasons have been offered. One was that the mathematics literature was simply not regularly accessible to applied social scientists. Thus, researchers outside the circle remained ignorant of published and unpublished findings. An exemplary episode in this regard is reported by William J. McGill (1988) in his essay honoring George A. Miller, a mathematical psychologist at the center of the developing program of cognitive science at that time. McGill reports an encounter at the Institute for Mathematical Studies in New York where Miller presented a paper on the complete probability distribution discovered by him and McGill while working on free-recall learning. Upon finishing the presentation, a member of the audience (Professor Max Woodbury) announced that he had published these findings earlier in the *Annals of Mathematical Statistics*.[8]

A second reason is provided again by Aumann (1987) who describes the first decade and a half as a period of relative social and intellectual confinement based on high solidarity and strong norms of communication. The epistemic culture he portrays, and one corroborated by other autobiographical accounts, is like a socio-intellectual crucible of internal competition and cooperation coupled with indifference toward, if not disdain for, other social sciences. Informal exchange dominated. The technical work of the program was inextricably intertwined with social interactions and conflicts. Martin Shubik recalls it as a time for "just having fun," although he also recounts the intellectual and emotional intensity of competition among individuals with widely varying personalities. Morgenstern's status as a pariah among the traditional

7. See note 2 above.

8. Max Woodbury worked with John von Neumann at Princeton.

economists on the faculty at Princeton added further isolation and inward communication.[9]

An outsider's view of this community offers some validation of these internal accounts: "At midcentury I was enormously attracted to the game theorists. They seemed to be the wittiest among all the social science communities. They wrote with a sense of humor. The games they concocted for their players were fascinating. They were mean people. They were always trying to do one another in. And sometimes, as among those notorious prisoners, they were deadlocked, even with their fate depending on trust in one another" (Bernard 1990, 331). Jessie Bernard, a sociologist, was introduced to game theory by Sidney Siegel at Penn State. Her initial citations of Luce and Raiffa 1957 are among the earliest to appear in sociology journals. In a paper on social problems as problems of decision, she argues on behalf of the appeal of games of strategy for conceptualizing such social problems as criminal behavior (Bernard 1958); Hamlet's decision dilemma is offered up for literary appeal. But the content of her treatment of game theory is typical of most early applications: games are treated as metaphors or linguistic devices (Back 1963). The axiomatic structure of game theory is not applied. An ostensible reason for this "soft" application of game theory is apparent in Bernard's allusion to the yet undeveloped state of the *n*-person game as described by Luce and Raiffa (1957).[10]

Other reasons for the exclusiveness of this program during this period present themselves. This confinement was made possible by the generous sponsorship of the Office of Naval Research, the Rockefeller Foundation, and the RAND Corporation during this period. Financial support for the young mathematicians and economists freed them from dependence on outside sources of financial support. Furthermore, the applica-

9. Remarks made at the Conference on the History of Game Theory by Martin Shubik on 5 October 1990. Shubik refers specifically to episodes in which he and Lloyd Shapley were in direct competition with John Nash over early solutions to the limiting properties of games as the number of persons increases. Interpersonal rivalries were common.

10. Bernard's initial exposure to game theory as an epistemic culture is recounted in the following footnote. "At Penn State I took a course with Sidney Siegel on nonparametric measurement and managed to struggle through. But . . . when in a faculty seminar in the mathematics department one of the men put an equation on the board that traversed two walls of the classroom. . . . one member of the class raised his hand and pointed to one particular point on the long equation. The others . . . nodded their heads in agreement. Not a word was needed. This was clearly a kind of communication I could *never* master" (1990, 145 n. 18).

tions of developing game-theoretic models and the disputes surrounding them may have been sufficiently esoteric and politically onerous to encourage exclusiveness and privacy. Mirowski's recent (1991) examination of the applications of game-theoretic models to war-related research during this period, with reference to internecine disputes over the conduct and purpose of this research (Ellsberg 1956), may provide more explanation for this enclosure at that time.

Instead, interview and biographical and autobiographical accounts corroborate each other that the front developed and disseminated the program relatively exclusively and informally in the initial period, via personal contact and unpublished information exchange occurring in the contexts of meetings and workshops or in the process of institutional mobility among researchers moving from job to job. The institutional network within which this informal traffic occurred was centered principally at Princeton with a strong linkage based on resource and information exchange to the RAND Corporation in Santa Monica, California. The Institute of Mathematical Statistics in New York, the Department of Mathematics and the Bureau of Applied Social Research at Columbia University, and selected Army and Navy operations research offices in Bethesda-Washington were relatively tightly linked on the Eastern seaboard. The University of Chicago and the Graduate School at Carnegie-Mellon at Pittsburgh in the Midwest and Stanford University (its Institute for Mathematical Studies in the Behavioral Sciences under the direction of Patrick Suppes and the Center for Advanced Studies in the Social Sciences) along with RAND on the opposite coast were also major stops.

The Stanford stop on the circuit is especially notable in the intellectual and social biographies of these research programs. There at the Center for Advanced Studies and at the Institute for Mathematical Studies in the Behavioral Sciences, researchers from throughout the social sciences regularly met. Game theorists (for example, Luce, Shubik, and Raiffa), Lewinians (Guetzkow), learning psychologists (Estes), mathematicians (Tukey), and philosophers (Suppes) mingled with the financial support of the Social Science Research Council among other public and private sponsors. The workshops attended and sabbaticals spent at these sites are credited by these authors for being particularly influential on their own work. This listing does not exhaust the participants or the meeting network, especially as it ramified over time.

The characteristics of this research program at this time are similar to those described by Gerald Geison as common in the development of research schools (1981). His comparisons of several famous laboratories in the history of science—Liebig's chemistry laboratory in Giessen, Germany, Claude Bernard's physiological laboratory in France, and Michael Foster's laboratory in England—yield common features: highly focused research programs, charismatic leadership, social cohesion and strong informal ties, institutional power bases including (in-house) publication outlets, and effective recruitment and mentorship (Geison 1981). As such, research schools are institutionally and geographically differentiated scientific styles, "tastes," interpretations of core disciplinary ideas, and specializations.

At the risk of overdrawing the comparison, these characteristics warrant closer attention with respect to the organization and development of game theory. Program focus, social solidarity, and mentorship were well in evidence at Princeton. Leadership was also strong and effective. Von Neumann's professional reputation and personal power were strong and his wider sphere of influence beyond Princeton evident in the resources he was able to garner for institutional development (Ulam et al. 1969). However, he may be better considered the more central programmatic (intellectual) leader, particularly in the earliest period of the program's development. Morgenstern, on the other hand, acted more as an organizational leader and major advocate for game theory to economics and beyond, continuing in this role long after von Neumann turned to other intellectual pursuits. Articles by Morgenstern in such general outlets as *Scientific American* (1949a) and *Kyklos* (an international social science journal, 1949b) and other popularization efforts such as news releases and symposia presentations confirm his role in the program. The incubation period this kind of resource environment and leadership provided has its precursors in nineteenth-century research schools in diverse scientific fields.[11]

While such organization and focus can provide the social foundation for scientific change, according to Geison (1981), it can also lead to local resistance to new ideas, alternative methods, and anomalous research findings. Pinch's study (1977) of quantum mechanics in the 1950s illustrates the importance of technical culture for the process of scientific

11. Oskar Morgenstern to *Time-Life* in Zurich, August 1956, Box 41, subject file General Material, 1955–58, OMP. Morgenstern repeats his oft-stated argument calling for a new mathematics appropriate to the subject matter of economics and credits von Neumann for providing the theory of games of strategy as a foundation for such a mathematics. This is another example of his popularization efforts on behalf of the program and von Neumann.

diffusion. He examines the "failure of communication" between David Bohm and John von Neumann and traces it directly to their distinct positions on axiomatization in quantum theory. Following Niels Bohr and the Copenhagen School in quantum physics, David Bohm argued in a paper in 1952 that different experimental apparatuses in quantum mechanics were associated with the distribution of hidden parameters that were crucial to the interpretation of quantum theory. This argument challenged a long-established von Neumann proof (published in 1932) that such "hidden variables" could not exist. Von Neumann's rigorous discipline of mathematical proof in which any theorem was internally derived from a few axioms and not from external or additional information reflected the mathematical ideal and clearly separated logical and empirical systems of knowledge. Today the proof is considered invalid (Bohm 1974). However, at the time of the Bohm proposal the hegemony of axiomatization and the arithmetic ideal (termed axiomania by Feyerabend 1969, 87) was epitomized in von Neumann's work and in the core rationality of game theory. This epistemology has been characterized specifically by Bohm (1974) as having produced habits of thinking and doing scientific work by those in von Neumann's tradition that resisted new ideas and practices.

Even within the international community of game theorists, competing schools of thought existed. Debates took place between the "American" and "French" schools of game theory. The French school initially adopted a more psychological approach to rationality. Their primary assumption was that there existed a "subjective distortion of objective probability" by individuals. Thus, such "human" variables as pleasures in gambling or superstition could lead individual evaluations of stimuli events to be quite different from what they actually were. In many respects, the American school was not seriously challenged by this difference in the definition of the problem domain. However, the presence of multiple schools of thought embedded in different cultural and organizational contexts lends credibility to the argument that local, technical cultures are important for the epistemic development of scientific research programs and for scientific development.[12]

12. Pierre L'Huillier to Morgenstern, paper titled "Utility Measures in Situations Involving Risk: A Few Reflections on the Problem by the 'French School' " (a summary of the International Colloquium on Econometry, Paris, May 1952), Box 47, OMP.

Coexistent Research Fronts

The limited diffusion of the theoretical core of game theory in this period is not accounted for exclusively by the behavior described above. Rather, the extent to which other fronts were developing their own programs under varying conditions of support and conflict also accounts for the diffusion of these ideas at that time. Indeed, the transepistemic network in which the game theorists were embedded intellectually and socially also housed other research programs. Moreover, some social scientists moved between these programs facilitating, but not always effecting, interprogram influence.

Some individuals on this landscape emerge as potentially critical disseminators. A citation analysis of the literature in the social sciences between 1956 and 1965 suggests that R. Duncan Luce's collaboration with Howard Raiffa in the publication of *Games and Decisions* (1957) was probably the most influential event in the literature of that period for the spread of game-theoretic thinking by the early 1960s. While *The Theory of Games and Economic Behavior* is also highly cited in the social sciences, its citation is often more ceremonial or historical in its purpose. Luce and Raiffa 1957, however, is often being applied directly in problem analysis, especially in political science (see Riker 1992). Their development of optimal strategy is widely applied across disciplinary settings spanning political science, education, law, philosophy, psychology, and sociology. The citation of Bernard 1958 earlier is illustrative of this pattern. Other early influences of this disseminating work outside of political science are evident in studies ranging from community structure (Long 1958) to deviant behavior (Cohen 1959) to behavior in small groups (Thibaut and Kelley 1959).

Besides being among the better-known game theorists in the wider social science literature, R. Duncan Luce also moved geographically among MIT (1942–45, 1946–53), Columbia (1953–57), Harvard (1957–59), and the University of Pennsylvania (1959–69) during this formative period in the front, working with social scientists in several research programs including the Lewinians and the survey researchers under Paul F. Lazarsfeld's direction at Columbia.

The Lewinians

Kurt Lewin, an experimental psychologist who worked in Berlin in the 1920s, settled in the United States in the mid–1930s (Mandler and Mandler 1969). By 1935 he was established at the University of Iowa in the Child Welfare Research Station and was attracting students from throughout the country. This period would mark the establishment of group dynamics, the study of social influence processes in groups. During the war he was a consultant for the Office of Strategic Services advising the assessment staff on personnel selection criteria (OSS 1948), the Office of Naval Research, and the Public Health Service.

After the war, he moved to MIT with some of his students and founded the Research Center for Group Dynamics where he attempted to develop a field theory in the social sciences. Field theory undertook a topological psychology employing spatio-physical metaphors to describe human action, developing and applying such constructs as "life space," "paths," "gradients," and "regions." The central interest was in the action of social forces on individual behavior. Social structure was emphasized over sentiment or ideology. The mathematical specification of the theory was never completed. But the research program emerging around it was highly productive in experimental social psychology. Lewin's students—among them Dorwin Cartwright, Leon Festinger, Ronald Lippitt, Alex Bavelas, Harold Kelley, Stanley Schachter, Kurt Back, Morton Deutsch, John Thibaut, J. R. P. French, and Albert Pepitone—and colleagues at MIT (e.g., Harold Guetzkow) conducted extensive studies on group communication and influence processes and published voluminously by the standards of the discipline at that time (e.g., Festinger, Schachter, and Back 1950; Bavelas 1950; Guetzkow 1951; Kelley 1951; Cartwright and Zander 1953).

The primary methodology involved laboratory or (other social) experiments and social network (matrix algebra) analysis, for which R. Duncan Luce was recruited to act as "captive mathematician" while a mathematics student at MIT (Luce in Lindzey 1989). This approach was an extension of "sociometry," a research area earlier associated with J. L. Moreno whose workshops in New York were attended by the group and whose subsequent reaction to the group's success was less than sanguine. By the early 1950s Moreno resented the crowded presence of the MIT group in the major journals (e.g., *Human Relations*, *Journal of Abnormal and Social Psychology*) including his own (*Sociometry*) and accused Lewin's students of engaging in "astute and Machiavellian practices" to promote their careers (Moreno 1953, 103). The team research style of the Lewinians anticipated new forms of scientific production. Laboratory studies were relatively inexpensive, lent themselves to replication and fine-grained extension, generated a productivity yielding significant visibility for the program, and attracted many students (Mandler and Mandler 1969).

Thus, the social experiment and a team research approach were central elements of this technical culture. Social laboratories teeming with experimental subjects (often drawn from undergraduate student bodies) required a physical capital and different social capital than that observed in game theory. Multiple authorship was more frequent when compared with the game theorists and broader dissemination through more general journals (rather than in-house journals) was the norm.

The transepistemic appeal of Lewin's program also contributed to its success. Mandler and Mandler argue that Lewin's research interests fit the "social temper of American life in the twentieth century" (1969, 404). His experimental attack on social problems of influence and leadership and his interest in the dynamics of personality suited the pragmatic American tastes for democratic leadership and personal success and was supported generously by federal (military) agencies and private foundations (Rockefeller) with interests in group behavior (Mandler and Mandler 1969).

After Lewin's death, the Center for Group Dynamics under the direction of Leon Festinger moved to Michigan and maintained throughout the late 1950s and early 1960s a sustained influence on social psychology generally. Some of the later group-dynamics research incorporated aspects of game theory (Back 1963; Guetzkow 1959; Anderson and Moore 1960; Moore and Anderson 1962). For example, Thibaut and Kelley (1959) began examining reward-cost matrices which were conceived as rule structures in equilibrium. The purpose was to determine whether exchange processes could transform relations in two-person and, later, *n*-person groups (Thibaut and Kelley 1979). But problems with the quantification of subjective assessments of value brought the program to an end (Costanza 1990).

By the mid-1960s behavioral ecology became the primary domain for the application of game theory with clinical interventions using token economies. In social psychology generally, however, there was a "retreat" from group research and a discovery of the cognitive world, i.e., of the naive epistemology of the self. This shift toward the individual as agent also coincided with a concern regarding the potential ideological odiousness of certain laboratory study designs employing researcher strategies of social control, including those associated with game theory and obedience experiments (see especially Milgram's studies of obedience to authority [1961]).

Paul Lazarsfeld and the Bureau of Applied Social Research

The Bureau of Applied Social Research at Columbia University was the immediate offspring of the Office of Radio Research, a major research project at Princeton University funded by the Rockefeller Foundation in 1937 to study the effect of radio on American society (Gitlin 1978, Lazarsfeld 1969). The director appointed to this project was Paul F. Lazarsfeld, a recent émigré to the United States from Vienna. He had come from a major research project in Vienna studying unemployment; within a few years in the United States with the assistance of fellowship support he had established a network of colleagues and support structures to position him as a central social science figure.

Lazarsfeld's intellectual interests were in applied psychology, and he quickly found an affinity between these interests and those of marketing and political opinion polling. Within a year of assuming the directorship, Lazarsfeld's reputation for institution-building was mounting. His major resource contacts for contracts and research operations were centered in New York City and by the fall of 1938 the Office of Radio Research (ORR) had moved to Union Square.

The research program of this operation centered on the analysis of attitudes and behaviors of populations, though no coherent theoretical agenda developed successfully. Lazarsfeld became associated with the broad notion of "latent strategy." Here he combined his psychological interests in psychosocial motivations with a Durkheimian sense of underlying principles of cognitive and social organization that could be revealed empirically. "Latent structures" were conceptualized as unobservable underlying phenomena with social meaning, such as political attitudes or ethnocentrism, which could be observed only indirectly as manifest variables measured by individuals' responses to multiple structured questions. Lazarsfeld introduced the term "latent structure model" in 1950 as a matrical (tabular) method by which classes of observations were partitioned along parameters assessed in terms of conditional independence. His efforts to develop data reduction instrumentation and a tabular-partitioning analytic strategy reflected a core assumption regarding the roots of collective behavior and the need to eliminate spurious distractions of appearances. As such, instrumentation (for the measurement of responses and attitudes) and survey sampling design became the central features of Lazarsfeld's operation.

The latent structure approach was more empirically grounded than axiomatic. This directly distinguished it from game theory. It could be distinguished from Lewinian group dynamics by its approach to the empirical world; where the laboratory experiment was central to Lewin and his colleagues, the social survey and tabular methods of assessing interindependence among measured variables absorbed the Columbia approach.

The dominant utility of survey research was to "search for specific, measurable, short-term, individual, attitudinal and behavioral 'effects' of media content" (Gitlin 1985). The empirical frame of reference focusing on proximal social influences on attitudes and behaviors had a similar epistemic appeal in the broader American context as Lewin's experimental problem-solving. The seminal study that produced the single most influential theory of this program examined the personal influence process in the diffusion of opinions in Decatur, Illinois, in the mid 1940s. The "two-step flow of communication" argued that opinion leaders mediated the transmission of media information to individuals.

In the wake of the Decatur study, the next major project moved Lazarsfeld's operation into the academically respectable bureaucratic (though frequently contentious) confines of Columbia University by 1944–45; it was a national survey of political influence (entitled *The People's Choice*). This project included the theoretical and methodological ingredients for theory and research program development. The Bureau of Applied Social Research became nationally visible with the publication of this original study, and a stream of subsequent studies extended the findings, but a strong theoretical program never developed (Gitlin 1978).

At this juncture it is important to mention that Lazarsfeld shares the position of developing and expanding the role of large-scale quantitative survey research with Samuel A. Stouffer. Stouffer, affiliated with Harvard University following the war, led the collaboration on the four-volume *Studies in Social Psychology in World War II*; two of these volumes, entitled *The American Soldier* (1949), are a landmark in the emergence of "big social science." Princeton University Press, in a volume entitled *Putting Knowledge to Work* (1952), celebrating the tenth anniversary of its leadership under director Datus C. Smith, Jr., cited the two-volume set as the "mother lode" of social science publications of the postwar period. In the same breath, the press (which also published *Public Opinion Quarterly*, the primary journal of research on survey research) also cited *The Theory of Games and Economic Behavior* as contributing to the new importance of the social sciences in American society. Writing on this issue, J. Douglas Brown commented that the "university press is torn between the devilish temptation to go popular—in a field where every reader is his own expert—and the thrilling urge to sail the high seas of controversy" (Brown 1952, 14). His observation reflected an ambivalence toward the new social sciences whose proofs and observational techniques brought messages to the public regarding both the guile and vulnerabilities of human beings.

The Stouffer collaborations and the Bureau of Applied Social Research's analyses of American political attitudes and behaviors represented a move away from grand theorizing in the social sciences and toward tight empirical analyses of applied problems. The preference for empirical indicators of underlying (latent) social processes as well as the pursuit of proximal causes of attitudes and behaviors were departures from social science prior to World War II. And while these were distinct efforts to move toward quantification, they were nevertheless different strategies from that offered by game theory. They did not fit easily with axiomatic thinking.

However, the Bureau of Applied Social Research did become a setting for disseminating mathematical thinking in social research. A series of seminars met for several years over topics associated with quantification and formal analysis. Presentations by Rashevsky, Simon, Baumol, Vickery, and others were features of the seminars, some of which were published by Lazarsfeld in a collection (Lazarsfeld 1954). Game theorists were regular visitors at the bureau. Howard Raiffa was brought to Columbia by Lazarsfeld to participate in a program called the Behavioral Models Project which was established to introduce sociologists to such new approaches as Markov chains and game theory. Indeed, Luce and Raiffa wrote *Games and Decisions* (1957) while shuttling back and forth between the bureau at Columbia and Stanford. However, Raiffa reports that they wrote the book in four days together but never were called to assist anyone at the bureau with game theory.[13]

Lazarsfeld complained widely that his sociological colleagues resisted mathematical models of human behavior because of the association of these models with economics and, in turn, with business and the management sciences (1959). James Coleman, a leading contemporary exponent of rational theory in sociology today (1986, 1990a, 1990b) was trained at Columbia under Lazarsfeld's tutelage and reports his exposure to the mathematizing environment of that period and his own resistance to game theory. He writes,

> I did have some exposure . . . to game-theoretical ideas at the Bureau of Applied Social Research, both through exposure to von Neumann and Morgenstern's book, and because it was then and there that Duncan Luce and Howard Raiffa were working on *Games and Decisions* . . . . I had some interaction with Herbert Simon, who was applying some ideas from economics to sociological problems. . . . Some of the tasks Lazarsfeld set me to work on involved exploration of the role that "utility" played in economic theory. . . . Yet, I did not see much value in carrying over the economist's paradigm of rational action into sociology. (Quoted in Swedberg 1990, 49)

In this context game theory did not find fertile ground. Coleman has argued that the interest in Durkheimian questions regarding collectivities and group sentiments precluded the adoption of game-theoretic thinking. Also, the analytic strategies of elaboration in tabular analysis that were applied at the bureau were used always to determine the environmental forces operating on popular opinion, voting behavior, or marketing patterns, not to model individual decision-making. The survey method did not fit obviously with game-theoretic concerns with smaller numbers and rational actors. Yet, he suggests that the intellectual environment nevertheless solidified the commitments of those exposed to quantifiable social science. His recent serious concern for incorporating rational models into social theory is not traced by him to that time, however (Coleman 1990a).

13. Remarks made by Howard Raiffa at the Conference on the History of Game Theory on 5 October 1990.

The Proof, the Experiment, and the Survey: Epistemic Cultures as Research Fronts

Lazarsfeld, like Lewin and von Neumann and Morgenstern, was a charismatic figure by all accounts and fostered high levels of solidarity and productivity among those working with him (Coleman 1990b, Gitlin 1978). But the alleged initial failure by Lazarsfeld to quickly transmit game theory and related mathematical models from economics and biology to the other social sciences is best considered in light of the competing epistemic cultures of the three programs. They were all relatively successful as entrepreneurial undertakings in the postwar period. Considerable resources from the outside resulted in notable research productivity. The social sciences, in their many forms, gained credibility and power. The bureau's development of survey research design may still be sociology's single most important contribution to the social sciences and to administrative institutions in the wider culture.

However, each research program strongly identified with a different technical approach, uniquely succeeded in establishing its niche in the social science landscape (or its market in the political economy of the sciences of the period—depending upon the metaphor of preference), and resisted, at least temporarily, the predominance of one over the other. Yet, by the 1960s the transmission of the game as one of several models for social behavior, with and without the mathematical heuristics, had occurred, though with short-lived impact. Its entry into sociology, particularly, reflects the influence of game theory as it converged with alternative and sometimes competing paradigms of social behavior.

Patterns of Influence in the Dissemination of Game Theory

This wide-angle perspective of the transepistemic landscape of the social sciences does not by itself provide us with a view of the operation of influence in the development of game theory in the social sciences. Research applying game theory outside the circle of core game theorists grew rapidly in political science, economics, and the management sciences. Some of these histories are presented in other chapters of this volume. I will briefly review two of the most visible studies in sociology. They typify the operation of influence (and resistance) discussed more generally in the earlier sections and schematically presented (see figure 1) as intellectual influence sequences.

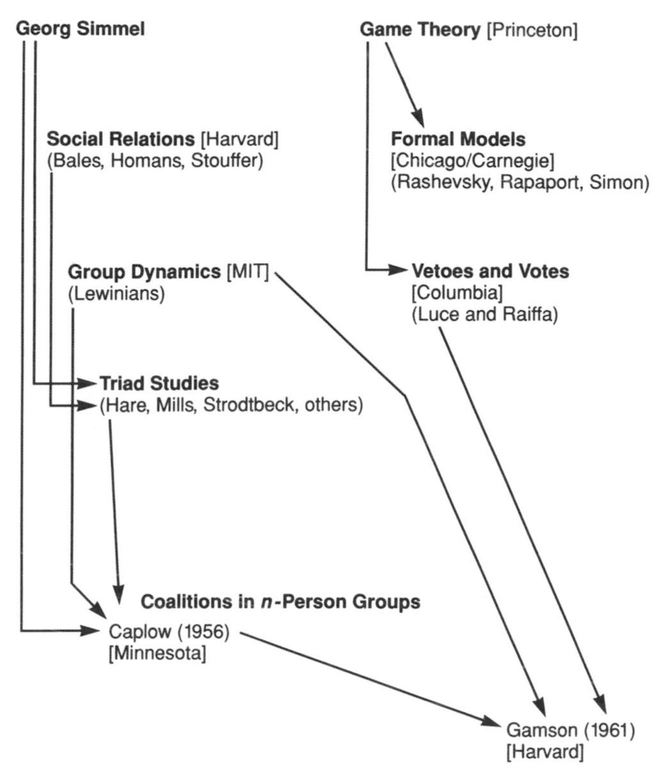

Figure 1 Case Study of the Diffusion of Game Theory into Sociology: Intellectual Influences on Coalition Research

Figure 1 portrays patterns of intellectual influence including collaboration, social contact, and direct citation across epistemic centers of social research in the postwar period. The game theorists, Lewinians, and Columbia researchers are joined in the influence system by mathematicians and formal theorists who exerted wide influence across the social sciences mentioned above (refer again to Kaplan 1956) and small-group and quantitative researchers falling within a Harvard circle of influence (the Social Relations group at Harvard and researchers on the social structure of the triad), with the latter representing a core sociological tradition extending back to Georg Simmel, a German sociologist from the turn of the century.

Theodore Caplow's (1956) theory of coalition processes in the triad was the first major effort in sociology to develop a critical extension of game theory by challenging the assumption of equality of power.

Caplow's fundamental argument was that explicit recognition of initial inequality in the distribution of power was sufficient to predict coalition processes. He formulated the thesis of "two against one" by arguing that a calculus of initial inequalities was a simple and more empirically valid alternative to axiomatic game theory for predicting the coalition alignment of individuals in three-person groups. But examination of his citations and discussions with him reveal that his interest in conducting the analysis came not from a direct concern for the von Neumann program but from consulting his colleagues at Minnesota at the time, specifically Stanley Schachter, an original member of the Lewinian group dynamics program at MIT. Caplow had been interested in Simmel's thesis regarding the significance of numbers as structural determinants of group formation and relationship behavior. He was not interested so much in the behavioral content of his social system but in the conceptualization of the influence of group structure, as the configuration of power relations, on individuals' behavior. Schachter, as a result of his work in the Lewinian program and his exposure to work done by Thibaut and Kelley (1959) on reward-outcome matrices, was sensitive to the game-theoretic program and introduced its developing ideas in the mid-1950s to Caplow.

Caplow approached his problem initially from a Simmelian standpoint. This approach was central to the work of other small-group researchers outside the Lewinian framework. Among this group, Bales (1950), Strodtbeck (Bales and Strodtbeck 1951), Homans (1958), Mills (1953), and others also attempted to quantify and matricize social interaction. Unlike Lewin, however, they did not attempt a social physics. Instead, they sought to link the imperatives of group structure and patterns of interaction derived from numbers with elementary notions of social exchange.

When Vinacke and Arkoff (1957) put Caplow's thesis of "two against one" to the test using pachisi board experiments, a brief spurt of research was initiated in the social psychology community using the same contrived game frameworks. This motivated Caplow to explicate more fully his theory and to turn to more substantive examinations coming from political science, anthropology, and economics. His motive was to attach the project to what he deemed more theoretically and substantively meaningful materials (Caplow 1968).

Caplow's developing perspective on his own application of game theory was not far from that described earlier as it emerged by the 1960s among the Lewinian researchers who applied these models (Thibaut and Kelley 1959). The small-group experiment was limited in assessing the intersubjective meanings that operate in natural relationships and situation. Like the Lewinians', Caplow's inclinations were to move toward substance (content). Upon doing so, his theoretical preferences outside game theory superseded his adoption of the mathematical program. His latest work on *Peace Games* (1989) bemoans the continued focus among game theorists on the prisoner's dilemma, though he cites with some enthusiasm Axelrod's (1984) work on the evolution of cooperation and the delineation of the rules of strategy produced by that work.

William A. Gamson's study of coalition formation published in 1962 is the second study in programmatic influence schematized (see figure 1). Gamson was at Harvard University when the study was published. He had been trained in the Michigan environment that included the Research Center for Group Dynamics founded by Lewinians and a growing circle of political scientists pursuing the interests recapitulated by William Riker (1992). And he acknowledges the direct assistance and support of Anatol Rapaport while he was conducting his research. Both influences are recognized by him as formative for his project. Unlike Caplow, he directly applied game-theoretic principles, using the Shapley-Shubik index and other mathematical heuristics of the program. And he managed, with the direct application of the program, to establish the inadequacy of the von Neuman–Morgenstern solution theory because of its profusion of solutions for many games. He applied constraints emanating from the program (Vickery's strong solution and the concept of psi-stability) to limit these results, but to no avail. And the Shapley value demonstrated, contrary to Caplow's predictions, the equal probabilities of coalitions despite the initial differences in resources. This led Gamson to the conclusion that while game theory worked effectively to improve the precision of Caplow's more verbal representations of power distributions, its prescriptive features—its profusion of solutions—made it ill-suited as a basis for descriptive theory.

The Caplow and the Gamson contributions and their respective precursors demonstrate the operation of influence across programs in a competitive landscape at various stages of development. The differential attraction and access to game-theoretic approaches is also demonstrated. The von Neumann–Morgenstern program, as handed down in the initial phase of the development of the program, slowed in its development and influence by the end of the 1950s. On the one hand, the relative insularity of the program restricted its visibility and accessibility to the wider landscape; on the other hand, even direct, personal exposure by game theorists to social scientists did not result in a conversion to game theory, or even (at Columbia) in communication. Its fundamental assumptions and heuristics, clearly recognizable even to the untrained, persisted in the 1960s to diminish its attractiveness among selected social science programs.

References

Abir-Am, Pnina G. 1982. How Scientists View Their Heroes: Some Remarks on the Mechanism of Myth Construction. *Journal of the History of Biology* 155.2:281–315.

Arrow, Kenneth K. 1951. *Social Choice and Individual Values*. New York: Wiley.

Anderson, A. R., and O. K. Moore. 1960. Autotelic Folk Models. *Sociological Quarterly* 1.4 (October): 203–10.

Aumann, R. J. 1987. Game Theory. In vol. 2 of *The New Palgrave: A Dictionary of Economics*. Edited by J. Eatwell, M. Milgate, and P. Newman. London: Macmillan.

Axelrod, Robert. 1984. *The Evolution of Cooperation*. New York: Basic Books.

Back, Kurt W. 1963. The Game and the Myth as Two Languages of Social Science. *Behavioral Science* 8 (January): 66–71.

Bales, Robert F. 1950. *Interaction Process Analysis: Method for the Study of Small Groups*. Cambridge, Mass.: Addison Wesley.

Bales, Robert F., and Fred L. Strodtbeck. 1951. Phases in Group Problem Solving. *Journal of Abnormal and Social Psychology*. 46.4 (October): 485–95.

Bavelas, Alex. 1950. Communication Patterns in Task Oriented Groups. *Journal of the Acoustical Society of America* 22.3:725–30.

Ben-David, Joseph, and Randall Collins. 1966. Social Factors in the Origins of a New Science: the Case of Psychology. *American Sociological Review* 31.4 (August): 451–65.

Berger, B. M., ed. 1990. *Authors of Their Own Lives: Intellectual Autobiographies of Twenty American Sociologists*. Berkeley: University of California Press.

Bernard, Jessie. 1958. Social Problems as Problems of Decision. *Social Problems* 6.3 (Winter): 212–20.

———. 1990. A Woman's Twentieth Century. In Berger 1990.

Bohm, David. 1974. Science as Perception-Communication. In *The Structure of Scientific Theories*, edited by F. Suppe. Urbana, Ill.: University of Illinois Press.

Brown, J. Douglas. 1952. In the Social Sciences. In *Putting Knowledge to Work*. Princeton: Princeton University Press.

Caplow, Theodore. 1956. A Theory of Coalitions in the Triad. *American Sociological Review* 21.4 (August): 489–93.

———. 1968. *Two Against One: Coalitions in the Triad*. Englewood Cliffs, N.J.: Prentice Hall.

———. 1989. *Peace Games*. Middletown, Conn.: Wesleyan University Press.

Cartwright, D., and A. F. Zander, eds. 1953. *Group Dynamics: Research and Theory*. Evanston, Ill.: Row Peterson.

Cohen, Albert K. 1959. The Study of Social Disorganization and Deviant Behavior. In *Sociology Today: Problems and Prospects*, edited by R. K. Merton, L. Broom, and L. S. Cottrell. New York: Basic Books.

Coleman, James S. 1964. *Introduction to Mathematical Sociology*. New York: Free Press of Glencoe.

———. 1986. Social Theory, Social Research, and a Theory of Action. *American Journal of Sociology* 91.6 (May): 1309–35.

———. 1990a. *Foundations of Social Theory*. Cambridge, Mass.: Belknap Press of Harvard University.

———. 1990b. Columbia in the 1950s. In Berger 1990.

Coleman, William. 1985. The Cognitive Basis of the Discipline: Claude Bernard on Physiology. *Isis* 76.281 (March): 49–70.

Costanza, Philip. 1990. Personal communication. 9–10 April.

Crane, Diana. 1972. *Invisible Colleges*. Chicago: University of Chicago Press.

Edge, David O., and Michael J. Mulkay. 1976. *Astronomy Transformed: The Emergence of Radio Astronomy in Britain*. New York: Wiley.

Ellsberg, Daniel. 1956. The Theory of the Reluctant Duellist. *American Economic Review* 46. (5 December): 909–23.

Festinger, Leon, Stanley Schachter, and Kurt Back. 1950. *Social Pressures in Informal Groups: A Study of the Human Factors in Housing*. New York: Harper.

Feyerabend, Paul K. 1969. On a Recent Critique of Complementarity: Part II. *Philosophy of Science* 36.1 (March): 82–105.

Fleming, Donald. 1969. Émigré Physicists and the Biological Revolution. In Fleming and Bailyn 1969.

Fleming, Donald, and B. Bailyn, eds. 1969. *The Intellectual Migration: Europe and America, 1930–1960*. Cambridge, Mass.: Belknap Press of Harvard University.

Fuller, Steve. 1989. *Philosophy of Science and Its Discontents*. Boulder, Colo.: Westview.

Gamson, William A. 1962. Theory of Coalition Formation. *American Sociological Review* 26.3:373–82.

Geison, Gerald. 1981. Scientific Change, Emerging Specialties, and Research Schools. *History of Science* 14.1:20–40.

Gitlin, Todd. 1978. Media Knowledge: The Dominant Paradigm. *Theory and Society* 6.2 (September): 205–53.

Guetzkow, Harold, ed. 1951. *Groups, Leadership, and Men*. Pittsburgh, Penn.: Carnegie.

———. 1959. The Use of Simulation in the Study of International Relations. *Behavioral Science* 4 (July): 183–91.

Harsanyi, John C. 1977. Advances in Understanding Rational Behavior. In *Founda-*

tional Problems in the Special Sciences, edited by R. E. Butts and J. Hintikka. Dordrecht: D. Reidel.

Homans, George. 1958. Social Behavior as Exchange. *American Journal of Sociology* 63.6:597–606.

Institute for Scientific Information. 1989. *Social Sciences Citation Index 1956–1965*. Philadelphia: Institute for Scientific Information.

Jaques Cattel Press. 1974. *American Men and Women of Science: Economics*. New York: R. R. Bowker.

Kaplan, Abraham. 1956. Sociology Learns the Language of Mathematics. In *The World of Mathematics*, vol. 2. Edited by James R. Newman. New York: Simon & Schuster.

Kelley, H. H. 1951. Communication in Experimentally Created Hierarchies. *Human Relations* 4.1 (February): 39–56.

Knorr Cetina, Karin D. 1982. Scientific Communities or Transepistemic Arenas of Research? A Critique of Quasi-Economic Models of Science. *Social Studies of Science* 12.1:101–30.

———. 1991. Epistemic Cultures: Forms of Reason in Science. *HOPE* 23.1:105–22.

Lakatos, Imre. 1978. *The Methodology of Scientific Research Programmes. Philosophical Papers*, vol. 1. Cambridge: Cambridge University Press.

Latour, Bruno. 1987. *Science in Action: How to Follow Scientists and Engineers through Society*. Cambridge, Mass.: Harvard University Press.

Laudan, Larry. 1977. *Progress and Its Problems*. Berkeley: University of California Press.

Lazarsfeld, Paul F. 1950. The Logical and Mathematical Foundation of Latent Structure Analysis. In *Measurement and Prediction*, edited by S. A. Stouffer et al. Princeton: Princeton University Press.

———. 1954. *Mathematical Thinking in the Social Sciences*. Glencoe, Ill.: Free Press.

———. 1959. Reflections on Business. *American Journal of Sociology* 65.1 (July): 1–31.

———. 1969. An Episode in the History of Social Research: A Memoir. In Fleming and Bailyn 1969.

Lewin, Kurt. 1947. Frontiers in Group Dynamics: Concept, Method, and Reality in Social Science: Social Equilibria and Social Change. *Human Relations* 1.1 (June): 5–41.

———. 1951. *Field Theory in Social Science*. New York: Harper.

Lindzey, Gardiner, ed. 1989. *A History of Psychology in Autobiography*, vol. 8. Stanford: Stanford University Press.

Long, Norton. 1958. The Local Community as an Ecology of Games. *American Journal of Sociology* 64.3 (November): 251–61.

Luce, R. Duncan, and H. Raiffa. 1957. *Games and Decisions*. New York: Wiley.

Mandler, Jean Matter, and George Mandler. 1969. The Diaspora of Experimental Psychology: The Gestaltists and Others. In Fleming and Bailyn 1969.

McGill, William J. 1988. George A. Miller and the Origins of Mathematical Psychology. In *The Making of Cognitive Science: Essays in Honor of George A. Miller*, edited by William Hurst. Cambridge: Cambridge University Press.

McKinsey, John C. C. 1952. *Introduction to the Theory of Games*. New York: McGraw-Hill.

Milgram, Stanley. 1961. Dynamics of Obedience: Experiments in Social Psychology. Mimeographed report. National Science Foundation.

Mills, Theodore M. 1953. Power Relations in Three-Person Groups. *American Sociological Review* 18.4 (August): 353–57.

Mirowski, Philip. 1991. When Games Grow Deadly Serious: The Military Influence on the Evolution of Game Theory. In *Economics and National Security: A History of Their Interaction*, edited by Craufurd D. Goodwin. Durham, N.C.: Duke University Press.

Moore, O. K., and A. R. Anderson. 1962. Some Aspects of Social Interaction. In *Mathematical Models in Small Group Processes*, edited by J. Criswell, H. Solomon, and P. Suppes. Stanford: Stanford University Press.

Moreno, J. L. 1951. *Sociometry: Experimental Method and the Science of Society*. New York: Beacon.

———. 1953. How Kurt Lewin's "Research Center for Group Dynamics" Started. *Sociometry* 16.1 (February): 101–4.

Morgenstern, Oskar. 1949a. The Theory of Games. *Scientific American* 180.5 (May): 22–29.

———. 1949b. Economics and the Theory of Games. *Kyklos* 3.4:294–308.

Morgenstern, Oskar. Oskar Morgenstern Papers. Special Collections Department, Duke University Library, Durham, North Carolina (OMP).

Mullins, Nicholas C. 1972. The Development of a Scientific Specialty: The Phage Group and the Origins of Molecular Biology. *Minerva* 10.1 (January): 51–82.

Nash, John. 1953. Two-Person Cooperative Games. *Econometrica* 21 (January): 128–40.

Office of Strategic Services (OSS). 1948. *The Assessment of Men*. New York: Rinehart & Company.

O'Rand, Angela M. 1986. Knowledge Form and Scientific Community: Early Experimental Biology and the Marine Biological Laboratory. *Yearbook of the Sociology of the Sciences* 10:183–202.

———. 1991. *Disciples of the Cell: Research Fronts and Traditions in Biological Science*. Unpublished manuscript. Duke University.

Pinch, Trevor. 1977. What Does a Proof Do if It Does Not Prove?: A Study of the Social Conditions and Metaphysical Divisions Leading to David Bohr and John von Neumann Failing to Communicate in Quantum Physics. In *The Social Production of Scientific Knowledge*, edited by E. Mendelson, P. Weingart, and R. Whitley. Dordrecht: Reidel.

Putting Knowledge to Work. 1952. A Tribute to Datus C. Smith, Jr., on the occasion of his tenth anniversary as Director of Princeton University Press. Princeton University Press.

Riker, William. 1992. The Entry of Game Theory into Political Science. *HOPE*, this issue.

Shapley, Lloyd, and M. Shubik. 1954. A Method for Evaluating the Distribution of Power in a Committee System. *American Political Science Review* 48.3 (September): 787–92.

Simmel, Georg. 1955. On the Significance of Numbers for Everyday Life. In *Small Groups*, edited by A. P. Hare, E. F. Borgatta, and R. F. Bales. New York: Knopf.

Simon, Herbert A. 1954. Some Strategic Considerations in the Construction of Social Science Models. In Lazarsfeld 1954.

———. 1955. A Behavioral Model of Rational Choice. *Quarterly Journal of Economics* 69.1 (February): 99–118.

———. 1956. Rational Choice and the Structure of the Environment. *Psychology Review* 69.1:99–118.

———. 1957. *Models of Man*. New York: Wiley.

Smith, Bruce L. 1966. *RAND Corporation: The Case Study of a Nonprofit Advisory Corporation*. Cambridge, Mass.: Harvard University Press.

Stouffer, Samuel A., et al. 1949. *The American Soldier*. Volumes 1 and 2 of *Studies in Social Psychology in World War II*. Princeton: Princeton University Press.

Strodtbeck, Fred L., and Paul Hare. 1954. Bibliography of Small Group Research, from 1940 to 1953. *Sociometry* 17.2 (May): 107–78.

Studer, Kenneth E., and Daryl E. Chubin. 1980. *The Cancer Mission: Social Contexts of Biomedical Research*. Beverly Hills, Cal.: Sage.

Suppes, Patrick. 1961. The Philosophical Relevance of Decision Theory. *Journal of Philosophy* 58.21 (October): 605–14.

Swedberg, Richard. 1990. *Economics and Sociology Redefining Their Boundaries: Conversations with Economists and Sociologists*. Princeton: Princeton University Press.

Thibaut, John W., and Harold H. Kelley. 1959. *The Social Psychology of Groups*. New York: John Wiley.

Ulam, S., H. W. Kuhn, A. W. Tucker, and Claude E. Shannon. 1969. John von Neumann, 1903–1957. In Fleming and Bailyn 1969.

Ulrich's International Periodicals Directory. [1963] 1984. The Bowker International Series Database. New York: R. R. Bowker.

Urban, Dieter. 1982. Mobility and the Growth of Science. *Social Studies of Science* 12.3 (August): 409–33.

Vinacke, W. E., and A. Arkoff. 1957. An Experimental Study of Coalitions in the Triad. *American Sociological Review* 22.4 (August): 406–15.

von Neumann, John. [1932] 1955. *Mathematical Foundations of Quantum Mechanics*, translated by Robert Beyer. Princeton: Princeton University Press.

von Neumann, John, and Oskar Morgenstern. [1944] 1947. *The Theory of Games and Economic Behavior*, 2d ed. Princeton: Princeton University Press.

Weintraub, E. Roy. 1985. Appraising General Equilibrium Analysis. *Economics and Philosophy* 1.1:23–37.

Whitley, Richard. 1984. *The Intellectual and Social Organization of the Sciences*. New York: Oxford University Press.

Woodbury, Max. 1989. Personal interview. April.

Part 3 Crossing Disciplinary Boundaries

The Entry of Game Theory into Political Science

William H. Riker

In this study of the transmission of ideas, my role is to report on how game-theoretic notions flowed from von Neumann and Morgenstern (1947) into political science. Since I had some part in that development, this article will, necessarily, read a bit like autobiography, but my hope is that the message will be mostly historical rather than personal.

So far as I can discover the first paper to discuss game theory in a political science journal was Lloyd S. Shapley and Martin Shubik, "A Method of Evaluating the Distribution of Power in a Committee System," which was based on Shapley (1953) and was published in 1954, in the *American Political Science Review*. This journal, which is the organ of the American Political Science Association, is usually thought to be the main journal in the field. So the first game-theoretic paper in political science was launched in the channel likely to spread it about most widely.

I remember vividly how impressed I was at my first reading of that paper. Since I subsequently became something of a publicist for it, I suppose my state of mind at that first reading is relevant to this history. Though I use the personal pronoun, I think that my state of mind is representative of others (e.g., Herbert Simon, as represented in his book *Models of Man* [1957], or David Easton, as represented in his book *The Political System* [1953]) who were at that time looking for methods to improve political science.

The previous year I had published a textbook, *Democracy in the United States* (1953). Although I was proud that I had written for the intellectually upper end of the vast market for textbooks on American government, I was nevertheless disappointed as I looked over the product of my venture. Initially, I had believed that, by encasing the usual stylized facts inside a more or less coherent theory about the interdependence of governmental institutions and widely held individual goals, I would be able to explain both the institutions and the goals in a way that was both more memorable for students and more appropriate for scientific discourse. Perhaps I achieved my first intent since, with only one revision, it remained in print for twenty-five years and was translated into four other languages. Furthermore, in a review in 1964 of the textbooks of the previous thirty years, Theodore Lowi praised it for interpreting institutions in terms of goals, and he used words that a careless reader might believe meant it was the model work in his sample (Lowi 1964). But the book came nowhere near my goal of scientific viability, and, as a consequence, I was rather disappointed in it and rather depressed about the possibility of a genuinely scientific political science.

By 1954 I had diagnosed the problem for political science as one of theorizing. What passed as "political theory" in those days was simply some random normative prescriptions about the good society, with hardly any mention of institutional arrangements. This was what I had studied at Harvard in graduate school, and my program for improving on that kind of philosophizing was to introduce coherence by straightening out the relation between normative and descriptive sentences. But while the so-called political theory of the era was confused and irrelevant, that feature was not, of course, the reason it failed scientifically. Rather, the difficulty was that the theory was almost exclusively deontological, conceived and often written in the imperative mood. There were no testable sentences in the indicative mood about institutions or behavior. It is true that at the University of Chicago under the leadership of Charles Merriam, several political scientists (e.g., Harold Gosnell, Hermann Prichett, Herbert Simon, and V. O. Key) had tried to build quantitatively testable models. But Robert Hutchins, that dark and evil figure in American social science, had systematically driven them out of the university in favor of narrowly normative political philosophers such as Leo Strauss. Consequently, in the 1950s there was no intellectual center in political science where it was clearly understood what a science of politics was all about.

This, then, was my state of mind in 1954. Despite my poor education in social science, I had begun to understand that, to be scientific, political science needed testable models about political phenomena—that is, refutable descriptive sentences. But no such theory existed. There were a few hints: Maurice Duverger had proffered what is now called Duverger's law about the relation between plurality voting and the two-party system, although I had not then seen Duverger's work and knew about the proposition of the law from the writings of Arthur MacMahon and E. E. Schattschneider.[1] But aside from this so-called law and its rationalization with a nascent theory, there was very little empirical generalization and almost no theory. Because of this fact, I had been looking, somewhat randomly to be sure, for methods of constructing theory. I had looked, fruitlessly, at modern logic, and it was just at this juncture that I read the Shapley and Shubik paper and Kenneth Arrow, *Social Choice and Individual Values* (1951). These two works led me back to von Neumann and Morgenstern, *The Theory of Games and Economic Behavior* (1947). There I discovered what I thought that political science needed for constructing theory.

I was impressed by several features of game theory that fitted it into the tradition of political science. One feature was its uncompromising rationalism. Though it was normative in the sense that it permitted a theorist to recommend choices, still it based its analysis on what a goal-oriented rational chooser would choose. There was no role for instinct, for thoughtless habit, for unconscious self-defeating desire, or for some metaphysical and exogenous will. Rather, game theory analyzed social outcomes in terms of the interaction of participants, each of whom calculated to achieve some self-identified goals. Of course, it was not assumed that everyone, or even anyone, succeeded, merely that each chose with gain in mind and that the outcome was determined by the way all participants' choices affected each other. Most writers in the long tradition of interpreting politics had discussed events in terms of sensible people seeking to achieve rather straightforward and easily understood goals. Only in the first half of this century did biological, psychological, and metaphysical theories about instincts, drives, and "wills" challenge and nearly upset the traditional rationalism. My intuition was that people on the whole knew what they wanted and had pretty good ideas about how to get it. And so I was very impressed by the opportunity game theory

1. Later I traced its history back even further to Henry Droop in 1869, though I am sure it goes back even beyond that (Riker 1982). It is an interesting commentary on American political science in the 1950s that it took a French scholar to recognize the scientific potentiality of the one and only proposition that, in a conversation with me in 1955, Sam Beer could adduce as evidence of a political science. Fortunately, this sentence has subsequently been much studied, quantitatively by Rae (1967) and then by many others (Riker 1982) and recently theoretically by Cox (1987), Palfrey (1989), and Feddersen, Sened, and Wright (1990). This latter paper is based on noncooperative game theory so it is a fitting culmination of the effort to use game theory to generate political theory.

offered to organize investigation into rational choice and perhaps to put the results of the investigation into testable sentences.

A second feature of game theory that attracted me was its emphasis on free choice. Though determinism is a very old idea, it flourished anew in the nineteenth and twentieth centuries, turning humans into mechanistic agents of the forces of history or simply the carrier of some exogenously determined drive. Economic determinism, especially Marxism, and historical determinism of an idealistic (Hegelian) sort, as well as the previously mentioned irrationalisms came to dominate social thought. Game theory, however, allowed for free choice. It recommended that participants, knowing their own preferences, estimate how alternative strategies might satisfy those preferences in the face of similar calculations by opponents. It recommended then that participants choose strategies accordingly. Thus outcomes would depend on individual choices and the interaction among them—not on some exogenous plan for the world, not on some built-in irrational propensity, just on free human choice. This was welcome indeed to someone like me who believed in humanistic values.

One of the main reasons that deterministic assumptions attract students of politics is that these assumptions guarantee regularity of behavior—and thus admit generalization. Free-will assumptions, on the other hand, seem to imply random behavior that defies generalization entirely. This contrast confounds the humanistic social scientist. To generalize, the scientist needs behavioral regularities, which apparently prohibit human choice; but to respect human choice the scientist needs free will, which apparently prohibits generalization. Game theory offers a way out of this dilemma by combining the possibilities of generalization and free choice.

While game theory was initially presented as normative, as recommendations for behavior for those seeking to maximize against a similarly maximizing opponent(s), it was possible—and soon became customary—to treat it as descriptive by assuming that persons placed in conflict situations do indeed wish to so maximize.

As a descriptive theory, then, people with similar goals placed in similar circumstances are expected to choose among similar alternatives in approximately the same way. If, on examination of the world, it turns out that they behave as the theory describes, then, of course, the theorist has the desired scientific explanation of a class of natural events. If, however, people do not so behave, the theorist must revise his or her analysis because the failure to describe means that some feature of the situation has been misspecified. Typically, of course, it is the participants' goals that are misspecified so the theorist must attribute alternative goals and then reanalyze interactions in the hope of describing behavior accurately.

This feature of game theory has been, I believe, its most important contribution to political science. It has allowed for generalization about human choice in a way that admits of more or less precise determination of the form of human goals in particular social circumstances. Many observers are prone to attribute their own goals to others or to attribute what, if they were ever to be in others' shoes, they hope or believe would be their own goals. Game theory helps to compensate for this error, while at the same time promoting generalization. A rationalistic, voluntaristic theory that promoted scientific discourse was welcome indeed in 1954 and I began to study it avidly.

Much as I was impressed by the Shapley-Shubik paper, it has not in the long run turned out to be of much use in political science. The promise in the paper was, however, considerable. It offered a way to measure the effect of constitutional provisions. Constitutions prescribe a routine for making decisions. For any given set of roles for participants, the prescribed routine presumably gives different roles different degrees of influence over the outcomes. Similarly, changing routines changes the influence of roles and, holding routines constant but changing the roles (as, for example, by changing voters' weights in a weighted voting system) changes their influence. What the Shapley value did was to set the total influence over outcomes at a fixed amount and then, for a particular routine and set of roles, to measure the influence of each role, expressing each role's influence in terms of its share of the fixed sum. Thus Shapley and Shubik created a measure of influence or a "power index." They used as an example a measurement of the relative power of legislators in a simple tricameral system similar to the tricameral legislature in the United States.

Subsequently their index was used in a number of different theoretical applications: (1) Mann and Shapley (1964) estimated the relative powers of states in the electoral college, which measure helped Brams's study of presidential campaigns (1978). (2) Shapley and I used it for the analysis of weighted voting systems, which in the 1960s were touted as a response to the Supreme Court's requirement of the "one-man, one-vote" standard in *Reynolds v. Sims* (377 U.S. 533, 1964). (That holding endangered the seats of state representatives, city councilmen, etc. with undersized constituencies, and one way they tried to save their seats was to reduce their relative voting weight. But the Shapley-Shubik index shows that power over outcomes and weight in voting are not closely related. Hence the device of weighted voting does not satisfy the *Reynolds v. Sims* requirement [Riker and Shapley 1968]). (3) Ordeshook and I (1973, 154–75) used the index to show that the increase of the number of small nation members on and the change in the decision rule of the Security Council of the United Nations had almost no effect on the relative power of members, although it was justified as a means of increasing the influence of small nations. It did, however, significantly increase the opportunity of delegates from small nations to hold attractive jobs. We made no effort to determine whether the justification was simply an error or was a cover-up of a grab for patronage. (4) Brams and I used the index to show that in the growth of proto-coalitions to winning size, there were points at which the weights of growing but less than winning coalitions were such that rational participants not in the coalition would find it profitable to jump on the band wagon. Thus we offered a theoretical explanation of the so-called band wagon effect (Brams and Riker 1972).

Despite these interesting applications, it is reasonable to wonder whether or not the Shapley solution is descriptive of power or influence in the real world. Most persons who have tried to analyze power have interpreted it as the ability of one person to make another person do something the other would not otherwise do. While I have deep reservations about this (and most other definitions of power [Riker 1964]), it is clear that Shapley's definition is quite different. It involves not the ability to control persons but the ability to control outcomes by means of being the pivot or the marginal person between winning and losing coalitions: the last-added member of a minimal winning coalition. In an effort to examine the question of whether or not the Shapley value measured anything that people might want, I studied legislators' changes in party membership in the French National Assembly for the years 1951 to 1955 (Riker 1959). For each legislator who shifted from one party to another, I calculated the power index of that person before and after the shift. Treating the parties as weighted voters, I calculated the power index of the shifting voter's party and assigned to that person $1/n$ of the party's power, where n is the number of members of the party. Subtracting each shifter's index before shifting from his or her post-shifting index to arrive at the gain in power from shifting, I hypothesized that, if the shifters were in fact seeking to increase their chances to pivot, the sum of the gains would be a positive number. In fact it was a tiny negative number, almost zero. Of course, I did not infer from this discovery that members ignored the advantage of pivoting. My result perhaps merely showed that in a large assembly (617 members), pivoting power is not easily identified. Nevertheless, my calculation did give me pause. I continued to use the index, but I remained skeptical about just what it measured.

Straffin's discovery (1977) of the assumption underlying the Shapley-Shubik index greatly sharpened my skepticism. By far the largest use of the index was to estimate the effect of changes in real-world voting systems, often for the purpose of testimony in lawsuits over weighted voting and districting. In connection with this use Banzhaf (1965) developed an alternative index based on a more or less arbitrary method of calculation, namely, the chance, over all possible combinations of n voters, for the indexed voter to be on the winning side. This was in sharp contrast to the Shapley-Shubik index which was based on an individualistic solution to n-person games, namely the chance, over all permutations of n players, for the indexed player to occupy the marginal position between the winning and losing sides. But the mathematical niceties of the differences in the philosophical assumptions were not noticed and both indexes were widely used for testimony. The difference between the indexes was largely ignored because, on the basis simply of experience, it was assumed that the two indexes usually gave about the same results. However, Straffin, when calculating them both for a proposed constitution for Canada, found that they gave sharply different results in that case. In tracking down the reason, he discovered that Shapley and Shubik assumed a homogenous voting population in the sense that all voted yea with the same probability, while Banzhaf assumed independence or that all voted yea with a probability of one-half. These are, of course, vastly different assumptions, neither particularly reasonable, and it is never clear, when using these indexes for estimating long-term effects, which one of the two assumptions—or neither—is appropriate. This, of course, limits both the practical and theoretical use of any particular index of power.

Nevertheless, there remains the fundamental insight of power indexes: particular voters' influence over outcomes in a weighted voting system is not proportional to those voters' weights. Furthermore, various decision rules may advantage or disadvantage weighted voters in nonobvious ways. However, this observation, which is in my opinion a significant discovery, long antedates game theory. Luther Martin (1788), in objecting to the unequal representation of states in the House of Representatives in the proposed United States Constitution, calculated a power index for the states (assuming all representatives from a state voted en bloc, as, of course, they might and indeed must whenever the House elects a President). He found that the index for large states was greater than their weight, while the index for small states was less, a fact which, of course, confirmed him in his hostility to the Constitution (Riker 1987). Martin's use of this calculation supports my belief that attempts to make a closer numerical examination of the disproportionality are not likely to be convincing. The reason is that any index is likely to omit important endogenous features of the system measured, which means that weighting all permutations or combinations equally is simply not a reasonable description of political reality.

The ultimate failure in political science of the Shapley-Shubik index is that it depends too much on the kind of probabilistic calculations based on the law of insufficient reason. It thereby fails to exploit the most valuable features of game theory, namely the choice of strategies. It assumes that the pivot has an advantage: to hog the payoff. That is, it describes the outcome of an unspecified process by which players become pivotal and, once pivotal, bargain to realize the putative advantage. Of course this assumption misses some kinds of situations that would be apparent in the extensive form (e.g., when potential pivots bid to maximally losing coalitions for the chance to be a pivot). This omission is not so serious, however, as the assumption that all permutations are equiprobable. Experience with long-lived constitutions (e.g., the Constitution of the United States) suggests, however, that some players have the chance to pivot much more frequently than expected. Fortunately the 1950s saw more promising applications of game-theoretic ideas.

By far the most important and interesting such application was Robin Farquharson's *Theory of Voting* (1969). Farquharson tells us in his preface that, while studying international politics at Brasenose College, Oxford, in 1953, he analyzed the vote in the United States Senate on the League of Nations. This led him to other works on voting (Black 1948, Arrow 1951) and thence to game theory. He saw the possibility of combining game theory and the theory of voting by means of the notion of strategic voting. Up to that time discussions of voting had mostly assumed that voters voted sincerely (i.e., in accord with their myopic true tastes). But Farquharson saw that voters could also vote strategically by voting against their myopic true tastes in one ballot of a series in order to achieve a final outcome more to their advantage than what they would obtain by sincere voting. In the 1980s and 1990s this remarkable insight has become the focus of studies of voting. Indeed most formal analyses of voting now simply assume that voting is a game and voters always vote strategically. Thus game theory and social choice theory have been fully combined. But the combination took a long time.

After 1953 Farquharson went on to perfect his undergraduate paper on voting at Nuffield College and offered it as a thesis for the D.Phil. degree in 1958. The examiners (R. Braithwaite and M. Dick) remarked: "Mr. Farquharson's thesis is on a type of subject little studied in British Universities [or, I add, anyplace else], partly because in terms of formal disciplines it forms a bridge between two faculties." They then recommended the degree because his work qualified "by the originality and excellence of its content" (Nuffield College files). Perhaps because of the limitations Braithwaite and Dick remarked on, Farquharson did not easily find a publisher for his work, even though it won the 1961 social science monograph prize from the American Academy of Arts and Sciences. I read it in 1962 for the Yale University Press and recommended it unreservedly for publication. Unfortunately Farquharson insisted on several three-color plates, an expense the press was unwilling to assume. Several years later Martin Shubik undertook to get Farquharson's book published. As a Yale professor Shubik carried more weight with the press than I did and Yale brought it out in 1969 with several multicolored plates but, unfortunately, many typographical errors. (These are identified in Niemi 1983.) Although Farquharson's work had some immediate impact, its method for finding equilibria was difficult to use. After McKelvey and Niemi (1978) published a simplified method, Farquharson's work became an irreplaceable part of all theoretical studies of voting—three decades, unfortunately, after it was first worked out.

Other interesting political applications of game theory were also developed in the mid-1950s. One such idea, which I continue to believe is promising, but which has so far come to naught, is Luce's notion of psi-stability. In its one interesting political application, he and Rogow (1956) estimated the effect of a partially disciplined two-party system (like ours) on the operation of a tricameral legislature (also like ours). They obtained several nonobvious results, notably the proposition that presidents might reasonably fear party majorities larger than two-thirds of both houses, regardless of which party (the president's or the other) had these majorities. Interesting as this one application was, the idea of psi-stability has not, so far as I know, been used again for the interpretation of politics. I think of this as a failure resulting from the absence of an audience, and I have encouraged several people to extend and complicate the analysis, now that we have larger computers. Perhaps this last theme may yet be recovered.

The game-theoretic themes I have mentioned so far are, of course, quite peripheral to game theory as a set of mathematical or economic ideas, at least as they were understood in the 1950s. In a sense, therefore, political scientists first glimpsed game theory through an open side door. This changed, however, with the publication of Luce and Raiffa's *Games and Decisions* (1957). This work contained an explanation of the von Neumann solution for two-person zero-sum games and of the von Neumann–Morgenstern solution for n-person cooperative games, as well as a description of a number of two-person non-zero-sum noncooperative games such as the prisoner's dilemma. I do not know how widespread was the use of this book in political science, but I do know I began to use it (in small classes) almost immediately and I think at least a few other people such as Glendon Schubert did also. Ultimately it had a significant impact on political science because it served as the undergraduate introduction to game theory for most of the political scientists who studied the subject in the 1960s and 1970s.

Of the three subjects Luce and Raiffa introduced, the von Neumann solution to two-person zero-sum games had the least influence. Aside from Brams's *Presidential Campaign Game* (1978), I can't think of any political research that used it in a serious way. The reason two-person zero-sum theory was something of a dead end, even though it was the first subject studied in most textbooks and courses on game theory, was that it was difficult for political scientists to see it as a model for any political process. This theory seems to have been attractive at the RAND Corporation because it was a good model for battles. But political scientists then were more interested in larger confrontational events, such as wars narrowly defined or wars defined in terms of their causes and consequences, and these larger events do not display either the zero-sum or the two-person features. Most wars are not zero-sum, that is, most leaders of a warring nation do not intend to annihilate the opposing nation or even the leaders of the opposing nation. Indeed, the most common war aim is to achieve some sort of accommodation between the opponents, an accommodation that is not zero-sum. Furthermore, most modern wars are not two-sided. Even if there are only two combatants, lurking in the background are allies and potential allies who have a significant effect on the outcome. So the two-person zero-sum model had little effect on the study of international politics.

It did have some effect on the study of elections which are, in the American context, mostly two-sided; and they are zero-sum in the sense that one candidate wins the office that is exactly what the other loses. The solution to games in this model is, however, a choice among strategies, and political scientists apparently could not simplify the strategies sufficiently to apply the solution. In most complicated applications, the appropriate choice is a mixed strategy and it is hard to envision a mixed strategy in an election. Furthermore, the median voter theorem, which had been proved by Black (1948, 1958) and popularized by Downs (1957), suggested a quite different strategic model, namely, convergence to the ideal point of the median voter. Thus, the only significant application to elections was by Brams (1978), as mentioned.

By contrast the feature of game theory that caught on most quickly in political science was two-person non-zero-sum noncooperative games. The prisoner's dilemma game had been studied by Deutsch (1958) with social-psychological applications in mind, and Rapoport and Chammah (1965) applied this game model directly to international politics. I will not follow the tortuous path of this model to cold war, nuclear exchange, etc. situations. The applications have been myriad and, I believe, illuminating as metaphors and for the revelation of the tendencies toward non-Pareto-optimal outcomes. But they have not been scientifically useful because they have ignored the fact that Pareto-optimal outcomes have *always* characterized the bipolar politics of the post–World War II era.

In the 1970s, however, another application of the prisoner's dilemma model did have a significant impact, not only on political science, but also on population biology. This influential application was the analysis of repeated prisoner's dilemma games by Axelrod (1984), Maynard-Smith (1982), and others. This line of thought has made the prisoner's dilemma more useful politically than it was previously, because it does account for Pareto-optimal outcomes in the real world. Since, however, this work is at present ongoing, I will not discuss it further.

A more immediately significant political application of the model of two-person non-zero-sum noncooperative games was that by Schelling (1963). While Schelling made little use of formal game theory, the model of these games led him to some striking sociological observations which continue to shape thinking about political strategy until today. Schelling's insight was that a player could, by effectively reducing his alternative strategies, signal that he would, for certain, choose a particular strategy. Such a signal would, of course, influence the other player then to choose his own strategy appropriately. If the first player selected well, he might thereby force the second to choose a strategy especially advantageous to the first player. The classic example is, for me, the driver in the game of chicken who ostentatiously removes the steering wheel and throws it out of his automobile, thereby signaling the other player that, to survive, he must concede and turn.

Just as Schelling exploited games in the two-person non-zero-sum category to derive not a solution but a sociological principle of commitment, so I exploited the von Neumann–Morgenstern model of cooperative n-person games to derive the size principle (Riker 1963). Cooperative solutions to n-person games concern the division of the gains from coalition formation among the coalition members, while the size principle concerns the number of members or weights of members in winning coalitions. In political situations analogous to constant-sum n-person games, participants with perfect and complete information form—so the principle asserts—minimal winning coalitions, that is, coalitions just large enough to win and no larger. This sentence was derived directly from the von Neumann–Morgenstern model, but I tempered it by an ad hoc information principle that the less perfect and less complete the participants' information, the more participants increase the size of winning coalitions.

While I applied the size principle to international alliances and political parties in the United States, it turned out that most of the subsequent political applications were to the coalitions of cabinet governments in multiparty legislatures. Many students of cabinet government had previously sought to generalize about the duration and ideological composition of governing coalitions. It had not, so far as I know, occurred to anyone that the sheer size of coalitions was a politically relevant feature. With the size principle as an inspiration—either as something to apply or as something to test—a number of scholars analyzed the size of cabinet coalitions in multiparty legislatures, mostly in Europe, but also in Japan and Israel (Leiserson [1968], Axelrod [1970], DeSwaan [1970], Taylor and Laver [1973], Browne and Franklin [1973], Dodd [1976], and Schofield and Laver [1985]). Aside from the work on the prisoner's dilemma, this work on cabinet coalitions constitutes the largest application during the 1960s and 1970s of game-theoretic ideas to political analysis.

Of course, these researchers found that the principle was not a perfect predictor of coalition size, although Dodd (1976) pointed out it was a good predictor of the longevity of coalitions and, as Schofield pointed out, it was a slightly better predictor of size than were its alternatives. It has always been surprising to me that so many people thought that the principle might in fact always correctly predict the size of coalitions. The principle is derived from a very sparse model, one that requires a constant-sum condition, one that deliberately excludes ideology and tradition, one that myopically precludes agreements over time, and one that specifically assumes perfect and complete information which is seldom found in the natural world. Consequently, one should expect that natural coalitions only roughly correspond to the model. Instead, the usefulness of the model is that it reveals a significant constraint on coalition formation, not that it predicts the size of every natural coalition. Indeed, the remarkable fact, for me, is that this simple principle is frequently sufficient to explain coalitions, when so many other considerations might be involved.

One valuable consequence of the scholarly interest engendered by the size principle is that scholars have collected a large amount of data about coalitions. This has led them in the direction of trying to apply solution theory to interpret real world events. The von Neumann–Morgenstern solution is, I believe, too difficult to apply to highly complex situations, but in the 1970s and 1980s other more manageable solutions have been suggested and used by political scientists: the bargaining set (Aumann and Maschler 1964), the competitive solution (McKelvey and Ordeshook 1978), Schofield's solution (Laver and Schofield 1985), etc. So, finally, twenty years or so after the first use of cooperative game theory, scholars have begun to try out solution theories.

Unfortunately, this enterprise comes just at the time that, in economics, theorists are abandoning cooperative theory for noncooperative theory. So it may be that interest in noncooperative theory will lead to a reformulation of the problems of coalition theory. In some ways I think this may be premature. It is true that cabinet coalitions exactly fit the conditions for noncooperative theory rather than cooperative theory. Noncooperative theory assumes no enforcement, while cooperative theory assumes enforceable agreements. Cabinet coalitions, of course, concern the construction of the sovereign, that is, of the potential enforcer, so they are necessarily noncooperative. Still, cooperative theory seems to be relevant, especially in a foresighted world where reneging on agreements has nasty consequences. Implicit enforcement may be just as effective as explicit enforcement. In the mid-1960s I tested the von Neumann–Morgenstern cooperative solution to a three-person game by an experiment of about three hundred trials using college students as subjects. For three players the von Neumann–Morgenstern solution of this game was a set of imputations: {(0, 2.50, 3.50), (1.50, 0, 3.50) and all other imputations dividing an additional dollar among the three players and the experimenter, (1.50, 2.50, 0) and other imputations dividing two dollars}. The default outcome was (0, 0, 0) and the null hypothesis was the set of imputations {(0, 3.00, 3.00), (2.50, 0, 2.50), (2.00, 2.00, 0)}. In the three hundred outcomes there were about 10 percent defaults, but almost none approaching the null hypothesis. The average of all nondefault outcomes was very close to the principal points of the von Neumann–Morgenstern solution, although there was absolutely no enforcement procedure. The subjects wanted an enforcer and indeed often pleaded with the experimenter for him to serve as an enforcer. Of course he refused. Very imaginatively, the subjects invented enforcement devices, such as temporary trades of objects of value only to the initial owner. Of course, such trades did not really enforce, but the subjects, in their desperation, made them do so anyway. Consequently the subjects managed most of the time to come out near the cooperative solution, although they occasionally defaulted (Riker and Zavoina 1970).

I conclude from this experiment that the cooperative solution is attractive even in a noncooperative situation. Of course in my experiment it might benefit players and could not hurt them to try for the cooperative solution. This solution, with its internal and external stability, seemed attractive—every group of subjects recognized it and gave it a name even though they were unsophisticated about game theory. All this suggests to me that there is a strong force toward a cooperative solution, even in a noncooperative setting, when the game is repeated, as of course it is in the real world. So I would not advise researchers to give up on cooperative theory yet, even though the intellectual temper of the times favors noncooperative theory which is now being used in a wide variety of applications in political science.

References

Arrow, Kenneth. [1951] 1963. *Social Choice and Individual Values*, 2d ed. New York: Wiley.

Aumann, Robert J., and Michael Maschler. 1964. The Bargaining Set for Cooperative Games. In *Advances in Game Theory*, edited by M. Dresher, L. S. Shapley, and A. W. Tucker. Princeton: Princeton University Press.

Axelrod, Robert. 1970. *Conflict of Interest*. Chicago: Markham.

———. 1984. *The Evolution of Cooperation*. New York: Basic Books.

Banzhaf III, John F. 1965. Weighted Voting Doesn't Work: A Mathematical Analysis. *Rutgers Law Review* 19:317–43.

Black, Duncan. 1948. On the Rationale of Group Decision Making. *Journal of Political Economy* 56 (February): 23–34.

———. 1958. *The Theory of Committees and Elections*. Cambridge: Cambridge University Press.

Brams, Steven. 1978. *The Presidential Election Game*. New Haven: Yale University Press.

Browne, E. C., and M. N. Franklin. 1973. Aspects of Coalition Payoffs in European Parliamentary Democracies. *American Political Science Review* 67.2 (June): 453–69.

Cox, Gary. 1987. Electoral Equilibrium under Alternative Voting Institutions. *American Journal of Political Science* 31.1 (February): 82–108.

DeSwaan, Abraham. 1970. An Empirical Model of Coalition-Formation as an N-Person Game of Policy Distance Minimization. In *The Study of Coalition Behavior*, edited by Sven Groenning, E. Kelley, and Michael Leiserson. New York: Holt, Rinehart, and Winston.

Deutsch, Morton. 1958. Trust and Suspicion. *Journal of Conflict Resolution* 2.4 (December): 265–79.

Dodd, Lawrence C. 1976. *Coalitions in Parliamentary Government*. Princeton: Princeton University Press.

Downs, Anthony. 1957. *An Economic Theory of Politics*. New York: Harper & Brothers.

Easton, David. 1953. *The Political System: An Inquiry into the State of Political Science*. New York: Knopf.

Farquharson, Robin. 1969. *Theory of Voting*. New Haven: Yale University Press.

Feddersen, Timothy J., Itai Sened, and Stephen G. Wright. 1990. Rational Voting and Candidate Entry under Plurality Rule. *American Journal of Political Science* 34.4 (November): 1005–16.

Leiserson, Michael. 1968. Factions and Coalitions in One-Party Japan: An Interpretation Based on the Theory of Games. *American Political Science Review* 62.3 (September): 770–87.

Lowi, Theodore. 1964. American Government, 1933–1963: Fission and Confusion in Theory and Research. *American Political Science* 68:589–99.

Luce, R. Duncan, and Arnold A. Rogow. 1956. A Game Theoretic Analysis of Congressional Power Distributions for a Stable Two-Party System. *Behavioral Science* 1.2 (April): 83–95.

Luce, R. Duncan, and Howard Raiffa. 1957. *Games and Decisions*. New York: Wiley & Sons.

Mann, Irwin, and Lloyd S. Shapley. 1964. The A Priori Voting Strength of the Electoral College. In *Game Theory and Related Approaches to Social Behavior*, edited by Martin Shubik. New York: Wiley.

Maynard Smith, John. 1982. *Evolution and the Theory of Games*. Cambridge: Cambridge University Press.

Martin, Luther. [1788] 1981. *The Genuine Information, Delivered to the Legislature of the State of Maryland Relative to the Proceedings of the General Convention Lately Held at Philadelphia*. In *The Complete Anti-Federalist*, vol. 2, edited by Herbert Storing. Chicago: University of Chicago Press.

McKelvey, Richard D., and Richard G. Niemi. 1978. A Multistage Game Representation of Sophisticated Voting for Binary Procedures. *Journal of Economic Theory* 18.1:1–22.

Niemi, Richard G. 1983. An Exegesis of Farquharson's *Theory of Voting*. *Public Choice* 40:323–28.

Palfrey, Thomas R. 1989. A Mathematical Proof of Duverger's Law. In *Models of Strategic Choice in Politics*, edited by Peter C. Ordeshook. Ann Arbor: University of Michigan Press.

Rae, Douglas. [1967] 1971. *The Political Consequences of Electoral Laws*. New Haven: Yale University Press.

Rapoport, Anatol, and Albert M. Chammah. 1965. *Prisoner's Dilemmas: A Study in Conflict and Cooperation*. Ann Arbor: University of Michigan Press.

Riker, William H. 1953. *Democracy in the United States*. New York: Macmillan.

———. 1959. A Test of the Adequacy of the Power Index. *Behavioral Science* 4.2 (April): 120–31.

———. 1963. *The Theory of Political Coalitions*. New Haven: Yale University Press.

———. 1964. Some Ambiguities in the Notion of Power. *American Political Science Review* 58.2 (June): 341–49.

———. 1982. The Two-Party System and Duverger's Law: An Essay on the History of Political Science. *American Political Science Review* 76.4 (December): 753–66.

———. 1986. The First Power Index. *Social Choice and Welfare* 3.4:293–95.

Riker, William H., and Lloyd Shapley. 1968. Weighted Voting: A Mathematical Analysis for Instrumental Judgments. In *Representation: Nomos X*, edited by J. Roland Pennock and John W. Chapman. New York: Atherton.

Riker, William H., and Peter Ordeshook. 1973. *An Introduction to Positive Political Theory*. Englewood Cliffs, N.J.: Prentice Hall.

Riker, William H., and Steven J. Brams. 1972. Models of Coalition Formation in

Voting Bodies. In *Mathematical Applications in Political Science, VI*, edited by J. L. Bernd. Charlottesville: University Press of Virginia.

Riker, William H., and William James Zavoina. 1970. Rational Behavior in Politics: Evidence from a Three-Person Game. *American Political Science Review* 64:48–60.

Schelling, Thomas. 1963. *The Strategy of Conflict*. New York: Oxford University Press.

Schofield, Norman, and Michael Laver. 1985. Bargaining Theory and Portfolio Payoffs in European Coalition Governments 1945–1983. *British Journal of Political Science* 15.2 (April): 143–64.

Shapley, Lloyd S., and Martin Shubik. 1954. A Method of Evaluating the Distribution of Power in a Committee System. *American Political Science Review* 48.3 (September): 787–92.

Simon, Herbert. 1957. *Models of Man, Social and Rational: Mathematical Essays on Rational Human Behavior*. New York: Wiley.

Straffin, Philip D. 1977. Homogeneity, Independence, and Power Indices. *Public Choice* 30:107–18.

Taylor, Michael, and Michael Laver. 1973. Government Coalitions in Western Europe. *European Journal of Political Research* 1:205–48.

von Neumann, John, and Oskar Morgenstern. [1944] 1947. *The Theory of Games and Economic Behavior*, 2d ed. Princeton: Princeton University Press.

Operations Research and Game Theory: Early Connections

Robin E. Rider

Promises, Promises

In the realms of science and technology, new instruments and methods have long shone with the promise of unspecified but splendid discoveries, observations, and applications. Nearly four centuries ago Galileo, singing the praises of his telescope and his discovery of the Medici stars (in fact, Jupiter's moons), held out the promise of more benefits: "Perhaps other things, still more remarkable, will in time be discovered by me or by other observers with the aid of such an instrument" ([1610] 1957, 28). Nearer to our time, American physicist Ernest O. Lawrence, campaigning for funds to build a new and bigger cyclotron, foresaw "discoveries of a totally unexpected character and of tremendous importance" (quoted in Heilbron et al. 1981, 30); the popular press trumpeted the need for atom-smashers, certain that big-machine physics would yield impressive, if as yet unknown, advances. Arguments in favor of America's space program invoked similar promises of scientific discoveries and commercially viable spinoffs.

So too did operations research and game theory seize the imagination of experts and amateurs alike. Actual and anticipated benefits of operations research and game theory, as described in popular accounts and academic papers, shone with the enthusiasm of the immediate postwar era. The time was seen to be ripe, for instance, for the wider application of operations research (OR) in both industry and government;[1] indeed, proponents of OR envisioned its "great importance in many fields" (undated memo [1946], Box 2, "Institute Committee," Morse papers). The National Research Council Committee on Operations Research, established in the late 1940s to explore the potential peacetime

Archival research was supported in part by NSF Grant SES–8700287.

1. G. P. Wadsworth and A. A. Brown to N. McL. Sage, no date, in MIT, Institute Archives, Philip M. Morse Papers (hereafter cited as Morse Papers), Box 2, "Institute Committee."

use and expansion of OR, also argued from a successful war record to the applicability of OR techniques "throughout the whole military effort," in "business and industrial activities, and [in] those governmental activities which have an operational character" (draft pamphlet, Box 9, "NRC—Committee on O/R," Morse Papers). One report of wartime accomplishments of operations research was deliberately written in a way that would both explain OR methods and foster their use in other contexts. After declassification the report was published as *Methods of Operations Research* by Philip M. Morse and George E. Kimball (1951).

Similarly, later commentators on game theory observed that John von Neumann and Oskar Morgenstern's path-breaking book was written with the intent that "the motivation, the reasoning, and the conclusions of [game] theory" might be absorbed outside the boundaries of mathematical theory (Luce and Raiffa 1957, 3). As game theory "caught on" (Hurwicz 1955, 78) and "took fire" (Churchman et al. 1957, 519), ardor for the utility of its concepts and conclusions swept well beyond the military settings where it found favor early on. The number of new editions and reprintings of von Neumann and Morgenstern's book bore witness to the appeal and perceived relevance of game theory. A later textbook observed, not without envy, that "Only a very few scientific volumes as mathematical as [the *Theory of Games and Economic Behavior*] have aroused as much interest and general admiration" (Luce and Raiffa 1957, 3).

Surveying Common Ground

Both operations research and game theory drew practitioners from different fields and anticipated applications beyond usual disciplinary confines. Adepts and advocates of OR and game theory, self-conscious about the novelty, importance, and definition of their respective undertakings, sought to survey their domains through the time-honored academic practice of compiling bibliographies. Compilers of these works helped secure claims of innovation and versatility for operations research and game theory; their labors can help us mark out the ground common to operations research and game theory from World War II through the early 1960s.[2]

2. Compare the decision of the National Research Council Committee on Operations Research "to compile and edit an up-to-date bibliography in the field." "Report on ORC" (May 1953), National Academy of Sciences Archives (hereafter NAS Archives), "NRCPS: Com on OR. 1950–54. General."

A bibliography of game theory published in the last of four important volumes of *Contributions to the Theory of Games* took the story up through 1959. Its compilers, Dorothea M. Thompson and Gerald L. Thompson, caught more than a thousand items "making substantial use of game theory" (407–53). More than 170 journals were represented as well as books and published conference proceedings. Excluded for the most part were abstracts of talks and papers and most technical reports in ditto or mimeograph form. Despite these omissions Thompson and Thompson succeeded in demonstrating broad and growing interest in the concepts, methods, and conclusions of game theory. Not surprisingly, they found most interest in fields such as mathematics, statistics, economics, operations research, and management science. The *Journal of the Operations Research Society of America* (*JORSA*), for example, accounted for several dozen entries.

Bibliographies compiled from the point of view of operations research reinforce the impression that OR and game theory shared much territory. A list of references assembled by Vera Riley at the Operations Research Office at Johns Hopkins University in 1951 included the first edition (1944) of von Neumann and Morgenstern's *Theory of Games and Economic Behavior* (11); the book also figured in the preliminary edition of a massive bibliography on OR assembled by James H. Batchelor (1952) and in the bibliography appended to Morse and Kimball's *Methods of Operations Research*. Batchelor continued to collect bibliographical information throughout the 1950s and in 1959 published a thick volume of some four thousand citations covering the period up through 1957. The result of his work was "an imposing indication of the universal nature and wide range of disciplines drawn upon in operations research" (vii). Batchelor added a supplementary volume of more than twenty-five hundred citations in 1962, which brought the coverage up through 1959, and a third volume of nearly sixteen hundred entries to cover the period through 1960. The index to the first volume lists four score works about game theory, games, and gaming; the two supplements add another hundred, plus forty relevant entries.

One Arrow in the Quiver

Almost from the outset, British advocates of OR drew attention to its claim to methodological "inclusiveness": operations research was said to use numerous "special techniques" and models and to put "together many disciplines into one whole" (Editorial notes, 1951, 38). Experimental American courses beginning in 1950 organized material about operations research around these methods and their applications. Game theory was one arrow in the operations research quiver. MIT's team-taught operations research course in spring 1950, for example, included a unit on "operations problems involving game theory"; the organization of the course drew heavily on the wartime report later published as *Methods of Operations Research* (Morse and Kimball 1951).[3] Similarly, in *Operations Research for Management*, based on papers presented at an introductory OR seminar at Johns Hopkins in spring 1952, by then familiar methodological "favourites [like] game theory, symbolic logic, communication theory, and, of course, linear programming [made] their appearance" (Jessop 1956, 55). Case Institute of Technology provided another site for operations research education in the early 1950s, with a program that featured "technique-oriented courses: Production and Inventory Control, Linear Programming, Queueing Theory, Game Theory and Monte Carlo Methods, and Sampling Theory" (Ackoff 1957, 100).

To reinforce the lessons of wartime OR and to promote the training of new practitioners, some OR boosters took up textbook-writing. Their works contributed to the canon of techniques and models appropriate for use in OR. Philip Morse, tireless advocate of operations research, praised such labors for forcing "the recognition of similarities in areas hitherto considered unconnected" and illuminating "the broad utility of a number of research techniques and mathematical models" (1956, 5). He urged the exploitation of "[a]ny field of mathematics, any technique of measurement that will bring results" (6). Though Morse did not always share the vision of OR held by his colleagues at Case Institute, there is a certain congruence between Morse's views on the range of appropriate methods and models and the coverage of a textbook that grew out of discussion and experience at Case. The latter volume, an *Introduction to Operations Research* by C. West Churchman, Russell L. Ackoff, and E. Leonard Arnoff, won praise from a British reviewer as "the first attempt of a systematic description of O.R. methodology and techniques, obviously based on a very comprehensive survey of actual O.R. studies in the industrial field" (E.K. 1957, 221–22). The book emphasized a number of different models used in OR, prominent among them the game-theoretic approach.

3. Course description (spring 1950) and report (24 August 1950), in Morse Papers, Box 9, "NRC—Committee on O/R" (part 2). Compare Morse and Kimball 1951.

Common Tension

Game Theory

So that operations research might exploit the full range of available models and methods, Morse urged his colleagues in OR to keep "abreast of the latest developments" in game theory, statistics, communications theory, cybernetics, and many other fields (L. H. C. T. 1955, 18–19). As noted by Thompson and Thompson, many of the advances in game theory were presented in *JORSA*, the major OR journal; other important sources, including the *Naval Research Logistics Quarterly* and Project RAND reports, reflected the importance of military sponsorship for much of this work. Accounts of the "latest developments" in game theory expected considerable mathematical sophistication on the part of their readers. An article about a "multi-stage, multi-person business game," for example, called attention to "the many fascinating mathematical problems inevitably forced upon us in the course of our study" (Bellman et al. 1957, 472). Other relevant articles proved new theorems, investigated convergence, explored the geometrical significance of best strategies, demonstrated the existence of well-defined values for games, and examined the mathematical structure of specific games without necessarily making explicit their applications.

Though most studies stressed the "theory" in game theory, some *JORSA* articles also addressed the question of applications. The context for most of these papers was military. One such account, presented by T. E. Caywood and C. J. Thomas at the first meeting of Operations Research Society of America (ORSA) in 1953 and published in *JORSA* in 1955, opened with a description of efforts in the late 1940s to apply concepts of zero-sum two-person games in evaluating weapons systems. Much work of this sort on fighter-bomber duels was done at Project RAND, the Operations Evaluation Group at MIT, Aberdeen Proving Ground, and Project CHORE at the University of Chicago (institutional home for Caywood and Thomas). In the analysis of these duels, strategic options were matters of firing range or firing time; a mixed strategy in a one-round duel then specified the "probability distribution according to which this point of fire is randomly varied from one duel to another." The paper offered a simple numerical example and solved for the value of the game (1955, 402–11).[4]

4. A brief discussion of such duels appears in Morse 1954.

Because of the relative simplicity of models used in these papers (most often zero-sum two-person games), existing theory generally yielded a satisfactory mathematical solution, but this very simplicity also aroused suspicion as to the fit of the models with military reality. A few articles in early OR journals spoke directly to the issue of practical applications of game theory. In somewhat backhanded fashion, R. S. Beresford, major in the Royal Hussars, and M. H. Peston of the British Army Operational Research Group argued for the relevance of game theory to military operations by reinterpreting "an actual tactical problem" in game-theoretic terms. The setting was Malaya; the assignment was to protect a convoy of food trucks against Communist terrorists. The convoy escort had four strategies open; the terrorists, two. That the escort commander in this instance adopted "what is in reality a Theory of Games mixed strategy" led the authors to anticipate "no great difficulty in the further application of game-theory techniques to other minor tactical problems." The article was published "by permission of the Scientific Advisor to the Army Council" (1955, 175).

O. G. Haywood, Jr., likewise chose two actual, indeed "well-known battle examples"—one in the European theater in World War II, one in the Pacific—to make a point about the fit of game theory with the conservative doctrine of the Estimate of the Situation as inculcated throughout the U.S. military establishment. Haywood showed that military decisions taken concerning troop position and reconnaissance patterns in these historical examples were optimal in a game-theoretic sense. Haywood then deployed the examples "to demonstrate to the skeptical reader" that real military decisions could profitably be discussed "in terms of game theory" (1954, 382).[5]

Operations Research

Comments about the balance, or imbalance, between the mathematics of a mathematical model like game theory and its application in practice were paralleled by debates among proponents of operations research about theory versus application in OR literature. Philip Morse, concerned to avoid fruitless debate and dissension and eager to win adherents for the new discipline, advocated simultaneous progress on both

5. See also his RAND report RM-528, "Military Doctrine of Decision and the von Neumann Theory of Games" (1951), and the abridgement of his student thesis for the Air War College in the *Air University Quarterly Review* 4 (1950).

fronts—the mathematical and the practical—in operations research.[6] During his term in office as ORSA's first president, Morse laid out major tasks facing the field: "to develop analytic techniques and to broaden their range of application," and then, in ORSA meetings and publications, to make "new developments known to all" (Morse 1953, 164). In taking this approach Morse drew on his familiarity with the productive tension of experiment and theory in the physical sciences. He also made a virtue of necessity, since all research on operations came imbued with a "peculiar duality" of application and theory (Morse 1951).

It was not going to be easy to achieve Morse's goals. One participant in the first meeting of the ORSA publications committee (convened, it might be noted, after the journal began to appear) lamented the consequences of using "specialized jargon or advanced mathematics in a group of such diverse specialists." Tjalling C. Koopmans was more pointed in lowering expectations about the journal's readership (and ORSA's base of common interest): "at least 1/2 of the papers published must be understood by 1/2 of the membership of the Society" (ORSA Publications Committee Minutes, 11 June 1952).

As part of his assessment of OR methods in 1956, W. N. Jessop took a careful look at the first eleven issues of the ORSA journal. He found that slightly more than half the papers were theoretical in nature (of these 39, game theory accounted for 2; linear programming, 3; Monte Carlo methods, 3; queueing and traffic, 6; etc.). He considered "the placing of emphasis on mathematical methods and on highly abstract treatments of general situations" wrong-minded, because such publication policies deterred others—presumably those more interested in applications—from "placing their investigations on record" (54).[7] Jessop's objections to the excessively mathematical character of postwar OR literature were shared by P. M. S. Blackett and other British OR practitioners.[8] Some saw this as a characteristic difference between American and British brands of

6. Morse also thought the debate over the definition of OR wasted time that would be better spent on solidifying the mathematical foundations for the emerging discipline and applying OR to pressing problems of the day, both in the military and in industry. His comment as retiring president of ORSA—operations research is defined as being what the members of ORSA do—exemplifies a Popeye'd attitude: I am what I am!

7. The journal *Management Science*, issued by The Institute for Management Science, was subject to the same criticism by Jessop because its "contents so far strongly resemble those of J.O.R.S.A." (56). Max Woodbury made pointed comments on the disciplinary tensions that contributed to the founding of TIMS during the Duke conference on the history of game theory in 1990.

8. On Blackett's attitudes, see the comments of the editor of *Operational Research Quarterly* in 1974, as quoted in Lovell 1975, 69.

OR. Russell Ackoff of Case Institute called attention, for example, to the "methods-and-technique" orientation of American OR, and saw the British as shying away from "theoretical papers dealing with the mathematical exploration of some types of problem that arise in operations research" (1957, 89).[9] The difficulty, as even the broad-minded Morse was forced to admit, was that "those mathematicians are much more worried about having the correct words for everything than they are in getting an actual solution to an actual problem!"[10]

There were, however, structural reasons for a mathematical discourse largely devoid of practical details in OR. The initial review of the journal *Management Science* in the British journal *Operational Research Quarterly* noted with disappointment that "most of the articles are exercises in pure mathematics," since "workers in industrial or military operational research units are not using these journals to publish the results of their practical work" (B. H. P. R. 1956, 24). A frequent obstacle was getting the client's "permission to publish industrial Operations Research results," a problem the NRC Committee on Operations Research considered "especially severe."[11] In particular, American university programs in OR were trying to cultivate close connections with industry—witness the conference on OR in 1955 "sponsored privately and exclusively for representatives of companies" participating in the MIT Industrial Liaison Program. Program organizers encouraged the inclusion of details of specific practical applications of OR in conference presentations because the objective of the program was to "give participating companies advance information" concerning the institute's research; the "confidential nature of the various examples," however, meant that such papers could not be published as presented.[12] By virtue of his Navy and MIT affilia-

9. Cf. Morse's characterization of the development of *JORSA* (now renamed *Operations Research*), as growing "more like advanced mathematics texts, rather than [like the pages] of journals of physical science" ("ORSA Twenty-Five Years Later," Morse Papers, Box 21, "ORSA General," 4).

10. Morse to George Shortley, editor of *JORSA*, 16 October 1958, Morse Papers, Box 21, "Journal of ORSA." Of course, it must be remembered that the announcement of the MIT course on OR in 1950 referred to OR as a new field of applied mathematics (Morse Papers, Box 9, "NRC—Comm. on O/R" [part 2]).

11. "Minutes of the Committee on Operations Research. (First meeting of the full Committee) at the NAS-NRC Building, March 23, 1951" (NAS Archives, "NRCPS: Com on OR. 1951–1953. Mtgs: Minutes," 2). The same sentiment was expressed in minutes of the first meeting of the ORSA Publications Committee, 11 June 1952 (Morse Papers, Box 11, "ORSA Publications Committee").

12. V. A. Fulmer, MIT Industrial Liaison Officer, to Merrill M. Flood, director of the

tions, Morse was well aware of the academic hazards posed by "military or commercial security restrictions," and noted that such restrictions could make it "impossible to report specific applications of operations research" (1953, 164).

An emphasis on the techniques and models of OR, rather than on the specifics of military or industrial application, amounted to an end run around the problem. As Morse explained, "It should be much easier . . . to report developments in techniques, both theoretical and experimental, because such reports can be more easily stripped of the details requiring secrecy" (164). Likewise, battles won in wars past could be recast in mathematical form without serious compromise to national security, if still-relevant details of weapon effectiveness were excised. As a result, much of the published literature on OR (like that on game theory but in part for different reasons) put far more emphasis on mathematical theory than on actual application, in spite of rhetoric that promised great practical benefits to be derived from operations research and game theory.

Dual Approach

The development of linear programming, and especially its mathematical formulation, forged perhaps the closest connection between operations research and game theory in the 1940s and 1950s. Certainly a period of intense work on the simplex computational method for linear programming problems roughly coincided with the formative years of game theory. More important, a fundamental link between linear programming and game theory was the underlying mathematical theory of linear inequalities.[13] George B. Dantzig, who played a central role in developing the simplex method, noted, however, that systems of linear inequalities "excited virtually no interest until the advent of game theory in 1944 and linear programming in 1947" (1963, 20). Von Neumann, introduced to Dantzig and his research in 1947, was readily able to translate basic theorems from the realm of game theory into statements about linear inequalities. He, Dantzig, A. W. Tucker, Tucker's students, and a

---

Columbia Institute for Research in Management of Industrial Production, 1 June 1955 (Morse Papers, Box 4, "Industrial Liaison Conference on OR 1955").

13. On an early book-length study of game theory and linear programming, see Vajda 1956. For a historical sketch of the origins and influences of linear programming, see Dantzig 1963 (chap. 2).

team of researchers at RAND began to explore the theory of inequalities and the mathematical equivalence of game-theoretic and linear programming problems (Dantzig 1963; see also Shubik 1953). Both military and economic problems informed the initial development of linear programming, as was also true for game theory, but the industrial potential of linear programming assumed a certain prominence in the 1950s, especially with the successful importation of linear programming models into refinery planning and scheduling.[14]

Those doing operations research found in linear programming a congenial method and promptly incorporated it into the OR arsenal of models and techniques. In the first volume of *JORSA*, for instance, Morse called for fuller exploitation of linear programming in OR (1953, 160); only three years later W. N. Jessop, assessing the "most-written or most-talked about" OR methods, claimed that "most people would agree, I think, that linear programming is an easy winner in this respect" (1956, 51).

In the literature on linear programming, as in other aspects of OR and game theory, many works emphasized the mathematics at the expense of applications. The mathematical duality of game theory and linear programming and the ease with which linear programming problems yielded to game-theoretic analysis worked to reinforce this tendency (Shubik 1953, 152). Opponents of overmathematizing certainly found reason to complain about the character of many linear programming papers, while they paid a grudging sort of tribute to the mathematical significance of such research. Jessop is again quotable, this time on what he saw as a misplaced emphasis on theory: linear programming is "a subject so delightful to the pure mathematician that many papers appear to have had their origin in sheer exuberance unsullied by any thought of a factual situation" (1956, 51).[15]

Tracking Opinions

In a succession of introductory textbooks on OR, we can track opinions as to the importance and utility of game theory in operations research

14. See the unpublished interview with Howard Crandall, History of Science and Technology Program, Bancroft Library, University of California, Berkeley. Cf. the oil refining example in Bennion 1960 as cited in Batchelor 1959, vol. 3.

15. On a dispute over the utility of mathematical formalisms in inequality problems, see Klein 1955 and Charnes and Cooper 1955. Charnes and Cooper played important roles both in developing the theory of linear programming and in applying it in industrial contexts.

from the time that the "mathematical treatment of games 'took fire'" (Churchman et al. 1957, 519) until the recognition of OR as an established discipline. A volume intended to acquaint management with the virtues of OR, based on an introductory seminar at Johns Hopkins in 1952, included game theory as one of the "mathematical models . . . found useful in operations research" (Morse 1954, 102). There game theory was praised as "potentially a valuable addition to the body of methods available to the operations research worker," primarily because game theory is designed to handle problems "where the simple maximization principle cannot be applied" (Blackwell 1954, 252). Crucial wartime experiences lent credence to such arguments, since OR teams in World War II used game theory in working out tactics for submarine evasion of searching aircraft and air engagements between bomber and fighter (Morse 1954, 110–11; see also Morse and Kimball 1951, 102–9).

Operations Research for Management and the textbook on OR by Churchman et al. three years later both held out the promise of important applications of game theory in industry and business as well as the military, since of course competitive contexts are not confined to the battlefield. Morse's 1954 article in *Operations Research for Management* alluded to the use of game theory "to some extent, in deciding on the timing of an advertising campaign" (111); Churchman et al. watered down the enthusiasm by noting that "very little has been accomplished by way of applying the theory," at least to judge by published accounts. The Churchman text commented that military applications had been mentioned but not made public, and noted that other papers "explore the possibility of applying [game theory]" without dealing "with actual industrial applications" (1957, 557).[16]

Several difficulties impeded successful application of game theory in OR. Since the utilization of the game-theoretic model requires that strategic options and their implications be "accurately and exhaustively" expressed, Blackwell challenged OR "to describe such choices and consequences in a way that enables the solution of the game to be meaningful from the point of view of its application" (1954, 253). Others saw the crucial obstacles as residing in the mathematics itself. According to Churchman, Ackoff, and Arnoff, even simple games involve "extremely complex" mathematics, and thus far "only relatively simple

16. They also state, "This theory has received a good deal of attention in the recent literature but as yet it has found relatively few practical applications" (517).

gaming situations have been mathematized" (1957, 518). Morse, perhaps reasoning from experience in the physical sciences, saw such limitations as typical of mathematical models, but held out hope: "Most games are too complicated for the theory to analyze in detail *as yet*" (1954, 110; my emphasis). Morse also saw that there were advantages to be drawn even from studying simple models, because such models helped to clarify crucial aspects of more complicated games.

The exceptional case was linear programming, which had already found significant application in industrial optimization studies in the early 1950s, subject, of course, to the computational challenges posed by large and complex systems of linear inequalities. Here the "important formal relationship between linear programming and the theory of games" (Churchman et al., 1957, 557) meant that theoretical advances in one could be turned to advantage in the other, hopefully with consequences for the practical application of both.[17]

By the 1960s some of the hopes for the utility of game theory within OR had begun to tarnish. An OR textbook of the late sixties, like the critiques mentioned above, found fault with the mathematics itself: game theory had proved "capable of analyzing very simple competitive situations," but beyond them—in *n*-person games, for example—"existing theory" remained less than satisfactory (Hillier and Lieberman 1967, 278). Morse's early enthusiasm for game theory as OR's own branch of mathematics[18] thus appears overoptimistic; ironically, the mathematics had not yet shown itself powerful enough to handle "most competitive situations in industry and elsewhere" (267; echoing Churchman et al. 1957, 517–18). And yet OR textbooks and courses continued to present game theory as a method worth learning. J. D. Williams's assessment in *The Compleat Strategyst* of game theory's "greatest contribution" thus remained germane. In Williams's view, the game theory framework of pure and mixed strategies, with payoffs arranged in matrix form and the incorporation of probability distributions, offered "valuable orien-

17. See p. 519 on the impact of game theory "on the development of linear programming." A decade later another text referred to what was by then the "usual method" for solving larger games: transform the problem into one of linear programming and then use the simplex method and an electronic computer to generate the solution (Hillier and Lieberman 1967, 272).

18. Morse states that "operations research has its own branch of mathematics, called the Theory of Games, first developed by Von Neumann and since worked on intensively by Project Rand and other military operations research groups. Here the aspects of competition are specifically included in the mathematics" (manuscript of talk on "Physics and Operations Research" [27 December 1951], NAS Archives, "NRCPS: Com on OR. 1949–54: Supporting Data," 16).

tation" to those, like practitioners of operations research, "who must think about complicated conflict situations" (1954, 217; quoted in Churchman et al. 1957, 557–58).[19]

References

Ackoff, R. L. 1957. A Comparison of Operational Research in the U.S.A. and in Great Britain. *Operational Research Quarterly* 8:88–100.

B. H. P. R. 1956. Review of "Management Science, 1." *Operational Research Quarterly* 7.1:23–24.

Batchelor, James H. 1952. *Operations Research: A Preliminary Annotated Bibliography*. Cleveland: Case Institute of Technology.

———. 1959–63. *Operations Research: An Annotated Bibliography*, 3 vols. 2d ed. Saint Louis: Saint Louis University Press.

Bellman, Richard, et al. 1957. On the Construction of a Multi-Stage, Multi-Person Business Game. *Journal of the Operations Research Society of America* 5.4:469–503.

Bennion, Edward G. 1960. *Elementary Mathematics of Linear Programming and Game Theory*. East Lansing: Michigan State University.

Beresford, R. S., and M. H. Peston. 1955. A Mixed Strategy in Action. *Operational Research Quarterly* 6.4:173–75.

Blackwell, David H. 1954. Game Theory. In McCloskey and Trefethen 1954.

Caywood, T. E., and C. J. Thomas. 1955. Applications of Game Theory in Fighter versus Bomber Combat. *Journal of the Operations Research Society of America* 3.4:402–11.

Charnes, A., and W. W. Cooper. 1955. Such Solutions Are Very Little Solved. *Journal of the Operations Research Society of America* 3.3:345–46.

Churchman, C. West, Russell L. Ackoff, and E. Leonard Arnoff. 1957. *Introduction to Operations Research*. New York: Wiley & Sons.

Crandall, Howard. Unpublished oral history transcript. History of Science and Technology Program, Bancroft Library, University of California, Berkeley.

Dantzig, George B. 1963. *Linear Programming and Extensions*. Princeton: Princeton University Press.

E. K., 1957. Review of Churchman et al. Introduction to Operations Research. *Operations Research Quarterly* 8.4:221–22.

Editorial notes. 1951. Non-Military Applications of Operational Research. *Operational Research Quarterly* 2:37–38.

Galileo. [1610] 1957. The Starry Messenger. In *Discoveries and Opinions of Galileo*, translated by Stillman Drake. New York: Doubleday Anchor Books.

19. See also Hillier and Lieberman 1967: "Indeed, its formulation, logic and criteria are important conceptual tools [in OR]. Thus, the primary contribution of game theory has been its concepts rather than its formal application to solving real problems" (267).

Haywood, Jr., O. G. 1954. Military Decision and Game Theory. *Journal of the Operations Research Society of America* 2.4:365–85.

Heilbron, J. L., Robert W. Seidel, and Bruce R. Wheaton, eds. 1981. *Lawrence and His Laboratory: Nuclear Science at Berkeley 1931–1961*. Berkeley: Office for History of Science and Technology.

Hillier, Frederick S., and Gerald J. Lieberman. 1967. *Introduction to Operations Research*. San Francisco: Holden Day.

Hurwicz, Leonid. 1955. Game Theory and Decisions. *Scientific American* 192 (February): 78–83.

Jessop, W. N. 1956. Operational Research Methods: What Are They? *Operational Research Quarterly* 7.2:49–58.

Klein, Bertram. 1955. Direct Use of Extremal Principles in Solving Certain Optimizing Problems Involving Inequalities. *Journal of the Operations Research Society of America* 3.2:168–75.

L. H. C. T. 1955. Report on the International Statistical Institute (Rio, June 1955). *Operational Research Quarterly* 7.1 (June): 18–19.

Lawrence, Ernest O. 1940. Considerations on the Design of Large Accelerators. Lawrence Papers. Bancroft Library, University of California, Berkeley. Quoted in Heilbron et al. 1981.

Lovell, Sir Bernard. 1975. Patrick Maynard Stuart Blackett, Baron Blackett, of Chelsea. In *Biographical Memoirs of Fellows of the Royal Society*, vol. 21. London: Royal Society of London.

Luce, R. Duncan, and Howard Raiffa. 1957. *Games and Decisions*. New York: John Wiley & Sons.

McCloskey, Joseph F., and Florence N. Trefethen, eds. 1954. *Operations Research for Management*, vol. 1. Baltimore: Johns Hopkins University Press.

Morse, Philip M. Papers. Institute Archives, MIT, Cambridge, Massachusetts.

Morse, Philip M. 1951 (27 December). Manuscript of talk on Physics and Operations Research. PS: Com on OR. 1949–54: Supporting data. National Academy of Sciences Archives, Washington, D.C.

———. 1953. Trends in Operations Research. *Journal of the Operations Research Society of America* 1.4:159–65.

———. 1954. Progress in Operations Research. In McCloskey and Trefethen 1954.

———. 1956. Statistics and Operations Research. *Journal of the Operations Research Society of America* 4.1:1–18.

Morse, Philip M., and George E. Kimball. 1951. *Methods of Operations Research*, rev. ed. Cambridge, Mass.: MIT Press.

Riley, Vera. 1951. List of References on Operations Research. ORO Library Notes no. 2. NRCPS: Com on OR. 1950–54: General. National Academy of Sciences Archives, Washington, D.C.

Shubik, Martin. 1953. Game Theory and Operations Research (abstract). *Journal of the Operations Research Society of America* 1.3:152.

Thompson, Dorothea M., and Gerald L. Thompson. 1959. A Bibliography of Game

Theory. In *Contributions to the Theory of Games*, 4, edited by A. W. Tucker and R. D. Luce. Princeton: Princeton University Press.

Vajda, Steven. 1956. *The Theory of Games and Linear Programming*. New York: Wiley.

Williams, J. D. 1954. *The Compleat Strategyst, Being a Primer on the Theory of Games and Strategy*. New York: McGraw-Hill.

Game Theory and Experimental Economics: Beginnings and Early Influences

Vernon L. Smith

Introduction

My charge from Roy Weintraub in preparing this article was to do a work on game theory and experimentation that would be a history based on impressions from the early years, and not "just a survey." The result is a conversation about the development of experimental economics and its game-theoretic and other origins, based on personal impressions as well as the literature. As part of this discourse, I survey and discuss the content of some early works and how I believe they have influenced later experimental research down to the present.

Personal Impressions of Origins and Motivations, 1952–1962

My entry into experimental economics research, beginning in January 1956, was directly influenced by Edward Chamberlin in 1952 as an entering graduate student at Harvard. It was customary for entering students who had previous graduate training to audit several first- and second-year courses for a few meetings to help determine which, if any, would not need to be taken for credit because of the adequacy of one's prior training. I was in this category,[1] and, having taken an exceptionally thorough course in imperfect competition from Dick Howey at the University of Kansas, I sat in on Chamberlin's course at Harvard for only the first two meetings. In the first meeting the class participated in one of his (Chamberlin 1948) supply and demand "experiments." These

1. My undergraduate degree had been in physics and electrical engineering at Cal Tech. Subsequently I took a master's degree in economics at the University of Kansas to determine if I would continue in the study of economics. This led me to "shop around" among the offerings at Harvard to find out which courses I would not need to take.

classroom demonstrations were not received by the graduate students with any sense of importance. (We received no monetary motivation and viewed them as parlor games.) In fact they tended to be discredited; at the time, I tended to share this view, and, along with my peers, to dismiss them. Why? I think there were three reasons. (1) Chamberlin, we thought, used these demonstrations in unabashedly self-serving ways to show that perfect competition theory was unrealistic nonsense, and, by implication, that theory perforce must take the path developed under the rubric of monopolistic competition. Although probably few of us rejected Chamberlin's premise in our hearts, his motivations were deemed too blatantly "unscientific." (2) No one saw Chamberlin's exercises as real experiments in the sense that an experiment teaches you something important about the world, behavior, or the credibility of a theory. I would much later come to a personal understanding that this was not an accurate portrayal of Chamberlin's work, but I think this correctly describes the reaction my cohorts and I had at the time. Perhaps the best evidence for this interpretation is that although many generations of Harvard graduate students were exposed to Chamberlin's classroom exercise, apparently I was the only one of them who ultimately carried further the idea of doing experiments in economics. (3) Anyone who was familiar with what economists were doing circa 1952, and we were at the heart of learning about such things, would not fail to see that Chamberlin's experiments were a lone outlier in the methodological spectrum then current in economics.

A few years later, in connection with teaching introductory economics at Purdue (Smith 1981), I reconsidered Chamberlin's experiments more thoughtfully and reached a different conclusion. His experimental procedures were as follows: each buyer received a card defining that buyer's maximum buying (reservation) price for a single unit of the commodity; similarly each seller received a card with the seller's minimum selling price for a unit; all subjects circulated in the room; when a buyer encountered a seller they attempted to negotiate an exchange price by some unspecified and unstructured procedure; if they failed to contract, each would seek a new opposite and attempt to negotiate an exchange; as buyer/seller pairs made contracts they would turn in their cards to Chamberlin, report their exchange price, and he would post their price on the blackboard. Sometime later we would learn what had been the supply and demand schedule defined by the group's reservation prices, and what were the results of the experiment. Since the prices and total exchange quantity would not be near the equilibrium prediction, the important lesson we were to take home was that competitive theory was not supported. This set the stage for a course in monopolistic competition.

Upon reconsideration it struck me that the idea of designing experiments to test a proposition derived from economic theory was right. But Chamberlin's market contained rather primitive institutional features. Today we would call it a random-meetings economy: more-or-less simultaneous bilateral bargaining with no opportunity for the complete multilateral dissemination of information, and no opportunity to learn by repeated exchange through successive market trading periods. One of the conditions sometimes stated for "perfect" competition is that all participants have complete knowledge of all offers to buy and sell. This condition was implicit in Marshall's celebrated description of a corn exchange (also see Stigler 1968, 3:181).[2] It is also satisfied in each commodity or stock trading area (pit or post) on organized exchanges, and this was the basis for my first deviation from Chamberlin's procedures. The second was to organize the market environment as a demand (supply) flow per unit of time, a trading "day," with demand (supply) reinitialized at the beginning of each new trading day. An experiment consisted of several trading days in sequence. The hypothesis was Marshallian: there is a tendency for the competitive equilibrium to be approached over time, conditional upon the supply/demand flows remaining stationary for a sufficient period of time. My thought was to give the competitive hypothesis a best shot by using the trading rules in worldly institutions thought to be our most competitive. These were akin to the "double auction" rules used on the New York Stock Exchange although the term "double auction" was not used in my 1962 paper.[3] Under these rules any buyer could announce a bid to buy and any seller an offer to sell.

2. Earlier versions of this condition asserted that the participants had to have perfect knowledge of all values, costs, etc. (see Shubik 1959, 169–72 for a discussion). In contemporary game theory, knowledge of all payoffs and their utilities must be not only complete but also common, so that I know that you know that I know, etc. your commodity valuations, and even that knowledge of the model is common (Aumann 1987, 473). The equilibrium tendencies apparent in thousands of experiments conducted over the last three decades make it evident that such strong conditions are not necessary in repeated games for subjects to approximate classical competitive or game-theoretic noncooperative equilibria.

3. Apparently, my first use of "double auction" was in Smith 1976b (also see 1976a) where I cite Leffler and Farwell's (1963, 186ff) description of the double auction rules of trading on the New York Stock Exchange. I had used the latter and its earlier editions as a reference source on trading rules since the early fifties.

Contracts can occur either by a seller accepting a buyer's bid or a buyer accepting a seller's offer. Later implementations (not used in the experiments reported in my 1962 paper) by myself and other experimentalists used the standard improvement rule requiring any new bid or offer to narrow the standing bid-ask spread.

This experimental design was worked out late in the autumn of 1955, for implementation in one of my classes in the spring semester. The first experiment was conducted in January 1956. By 1960 I had completed the eleven experiments reported in my first experimental paper (Smith 1962). Contrary to my own expectations and the results reported in Chamberlin 1948, the "regular" experiments tended to converge quickly to the predicted equilibrium. The exceptions were attributed to important treatment effects. One experiment (Test 4) used a constant cost environment, and stabilized above the competitive equilibrium. This was due directly to the absence of monetary rewards: replication with cash rewards equal to each trader's realized surplus plus a five-cent "normal profit" for each transaction yielded all trades at the competitive equilibrium (Smith 1962, 121 n. 9) by period 3. This established clearly for me that monetary rewards mattered, and I never again conducted an experiment without using rewards that exceeded the opportunity cost of the typical subject. A second experiment (Test 8) that failed to converge was attributed to a change in "market organization." Instead of a double auction this market was one-sided; sellers were permitted to make offers, and buyers could not announce bids to buy but could only accept the offers of sellers. (The paper misleadingly associates these rules with retail markets which we now classify as "posted offer" markets [see Smith 1976b and Plott 1986].) This functioned to the disadvantage of sellers, suggesting the "curse" of auction rules requiring your side to be active in price-making.[4] This experiment was the first of many of my later studies showing that institutions matter. The next appeared in Smith 1964, but it would be more than a decade before I would fully appreciate what it means to say that institutions matter (Smith 1976b; Plott and Smith 1978). A final experiment (Test 11) converged more slowly than the others. This experiment allowed sellers to freely enter or exit, but at a sunk cost equal to their assigned reservation price. It supported the

4. This result is perhaps a close cousin to the results described below as the "curse of (payoff) information" (Schelling 1957; Siegel and Fouraker 1960; Camerer et al. 1989); the active side of the market reveals more information about their willingness to trade then the passive side (Smith 1964, 182).

idea that long-run adjustments are compounded of a series of short-run equilibria that impinge upon entry-exit decisions and the idea that the economist's concept of sunk cost was relevant to understanding market behavior.

The long period from research inception (January 1956) to publication (April 1962) calls for some discussion (also see Smith 1981, 372–73). The primary reason for the delay was that I was not ready for publication until 1962. I was doing no more than two to three experiments per year, mostly using my classes for subjects. (It was not until the 1970s that experimental economics became a vocation, rather than an avocation, for me.) I thought a lot about methodology and experimental technique. Through it all I was struggling with the question of whether what I was doing was economics. Not that I didn't think it was or should be, but the external feedback I received was mixed and confused—partly because I, as well as my audience, was confused as to exactly what I was up to. So I kept at it and ultimately was convinced that I had something to say; the final problem was to articulate it in a coherent and convincing way. Science is rhetoric—persuasion (although I would not have said so in 1962)—as much as it is theory, interpretation, and experiment. The first draft was submitted in March 1960 to Mary Jean Bowman, editor of the *Journal of Political Economy*. It was accepted, subject to revision, in October of that year by Harry Johnson, the new editor, who confessed that he had been one of the referees on the original draft. But by that time I had embarked upon an analysis of the various equilibrating hypotheses (Walrasian, excess rent, etc.) presented in sections 4 and 5 of the printed version. These results were incorporated into a completely revised draft and resubmitted in July 1961. The following 5 October Harry sent me two new referee reports—with the suggestion that their comments "seem to me to require further work and rewriting on your part. I am therefore returning the two copies of the manuscript to you, together with abstracts from the referee's comments to guide you. How you deal with them I shall leave to you." On 13 October I submitted a detailed rebuttal to both referees and suggested that "the(ir) main conclusion seems to be that I should have written a different paper based upon a different (real world?) set of experiments." Two days later Harry replied with the following.

> Thank you for your long and very careful letter of October 11. It was not my intention in any way to suggest that you should rewrite your article into the article the referees would have written had they

> written one. If you feel that your paper stands up against their criticisms, or that the revisions you suggest are the only concessions that you think it is necessary to make to them, just send the article to me as revised, and I will publish it. If the referees feel strongly enough to write up their criticism as comments, and these comments are worth publishing, you will have the right to rejoin. But somehow, I don't think that we will have much trouble that way.
>
> I look forward to receiving the manuscript with the excess rent section rewritten, the revision of the statement about the Marshallian hypothesis, and any other small changes you wish to make. I don't think it's necessary to include more material on the two experiments with monetary payoffs, especially as that is part of a separate study unit.
>
> Thank you again for your letter. I hope that the readers' comments have not annoyed you irreparably; I wanted to get comments from well-qualified people, and I must confess that I was rather put out at the tone of some of them. At any rate, I have learned something about my referees from this experience. As you may suspect, this is a very difficult job inasmuch as one has to keep evaluating everyone, including oneself.

Experimentalists today often complain that they get referee reports from nonexperimentalists that are oblique to the thrust of their papers, that raise naive methodological issues, or that argue that the predictions of theory are for important "real world" situations, not the laboratory. But the above experience shows that this has always been so, and it will probably continue to be so, partly because knowledge of experimental methodology requires one to be directly involved in all aspects of experimental research; reading about experiments is not enough. We still carry a large responsibility to explain ourselves to others. But even with common laboratory experience, different experimentalists may come to different understandings, so that discussions of the methodological and interpretational implications of theory/experiment are inescapable in answering the question, do you see what I see? Also, the empirical and methodological connections between field and laboratory require articulation and research directed to these ends. Gradually this has become clearer to increasing numbers of economists outside of experimental economics—thanks in no small part to the fact that experimentalists have confronted such critiques when they arise. Experimental papers have commonly dealt with questions of research motivation, implications, and the laboratory/field connection. The discipline of having to defend our methodology has been salubrious and, in my view, goes a long way to explain the accelerating success of experimental economics in the last decade.

The most significant event in the 1960s for my personal experimental research development was meeting Sidney Siegel in the autumn of 1961. I was a visiting professor in economics at Stanford; Sid was a Fellow at the Center for Advanced Study in the behavioral sciences. We met socially, discovered we were both doing "experimental economics," then got together at the center for a long meeting afterwards. Sid was an extraordinary experimental scientist; his published work does not fully convey the immensity of his knowledge of laboratory technique, his energy, and his enormous store of ideas that were yet to be implemented. He died of a heart attack in November 1961, at just the beginning of a brilliant career. I am convinced that his death impeded substantially the subsequent development of experimental economics, which did not really take off until the late 1970s. If he had lived, a major experimental initiative would have been sustained at another university. His work influenced much of my later thinking. At the time of his death much of the content of *Bargaining Behavior* (1963) was in the form of four Pennsylvania State research reports: "Bargaining Behavior I" (Fouraker, Siegel, and Harnett 1961), "Bargaining Behavior II" (Fouraker and Siegel 1961), "Bargaining Information and the Use of Threat" (Siegel and Harnett 1961), and "Oligopoly Bargaining: The Quantity Adjuster Models" (Fouraker, Shubik, and Siegel 1961). Larry Fouraker pulled it all together for final publication (Fouraker and Siegel 1963), complete with all instructions and data protocols. Sid Siegel's concern for the content of instructions, his insistence that instructions should be part of one's report, his practice of *always* motivating subjects with meaningful cash rewards, and much more are all evident in this work, in his other publications, and in the work of experimental economists today. One cannot, I believe, comprehend the early years of experimental economics in the United States without dealing with his and his coauthors' contributions. This is why I have devoted a section below to discussing it. After my encounter with Sid Siegel in 1961, I eventually read all of his papers and books. My impression is that his intellectual descendants are far more numerous in economics than in psychology.[5] Certainly, the idea that subjects should be paid salient re-

5. I once asked Amos Tversky, "Whatever happened to the tradition of Sidney Siegel in psychology?" His reply: "You're it!"

wards has, with many important exceptions, not prevailed in the study of decision behavior by psychologists (Smith 1991).

Other Impressions from the Early Years

I contacted many people—mostly early contributors to experimental economics—asking them questions and seeking their perceptions of the early development of experimental economics. Detailed responses were obtained from Martin Shubik, Herbert Simon, James Friedman, and Reinhard Selten. I have rearranged their comments somewhat so that they appear as direct responses to the questions I posed.

Martin Shubik

Martin Shubik has been a seminal participant in experimental economics, not only because of his early collaboration with Fouraker and Siegel on their work, which was a major influence in the founding of experimental economics, but also because of his contribution to what he has called "gaming." Although the distinction usually made between experimental gaming and experimental economics (Shubik 1975) is probably justified, I have always been terribly impressed by the extent to which Shubik and I, with different intellectual backgrounds and different experimental research interests, have experienced remarkably parallel learning. What you learn about institutions is very similar.

Q1. Generally, what are your impressions of the origins of the interest in experimental games/economics? Chamberlin 1948 seems to be the first game/market experiment.[6] Were you influenced by this?

A. I found out about Chamberlin years after gaming. As far as I can see his class exercise was an isolated event and influenced no one I know. Bill Parker (economic historian) actually played in Chamberlin's game. He notes that it was used only as a class exercise, not with any thought of experiment.

6. There were earlier precursory (and today often recognized) contributions to experimental economics, but they failed to stimulate follow-on research, and they were devoted to experimental preference measurement (Thurston 1931). Wallis and Friedman (1942) provided a famous negative statement on the possibility and value of experiment for economics. At that point people simply had no appreciation of the methodological range and potential for experimentation in economics; for example, the experimental study of market behavior is not even considered as a possibility.

Q2. Who were the principal contributors? These seem to include Merrill Flood and Anatol Rapoport (1950s) in addition to Fouraker and Siegel (1960, 1963), yourself, Austin Hoggatt (1959), Reinhard Selten (Sauermann and Selten 1960) and Jim Friedman (1963). But are there others? What did you know at the time about the 1952 Santa Monica conference on "The Design of Experiments in Decision Processes" which led to Thrall, et al., *Decision Processes* (1954).

A. I believe that Merrill Flood was far earlier and to a certain extent more important than Anatol in getting game theorists and psychologists originally started. Furthermore, one must remember that Bellman produced the large business game in 1957. This certainly influenced me [see Selten 1990 and his comments below].

The people at RAND more or less split into two—Merrill Flood and the other people involved in the Thrall book and the more social-scientist–oriented. Thus one had Helmer, Dalkey, Shapley, Paxon, Goldhamer, and Speier in one form or the other involved with COW (Cold War Game), and Flood, Shapley, and the others noted, and Thrall considering small, more or less formal, experimental games.

Earlier still was the group at Princeton—Shapley, Nash, myself, McCarthy, Hausner—1949–52—we were not directly considering experimental games—but were considering *playable games* to illustrate paradoxes and cute aspects of game theory. Thus, "so long, sucker" was cooked up in Fine Hall to show how one could make double-crossing a virtual necessity to win. I also invented a game called Kremlin (finished finally in 1958—unsuccessfully peddled to Parker Brothers) to illustrate problems in lack of trust and secrecy. I am fairly sure that it was from these sessions that I finally wrote down, many years later, the dollar auction as a way of illustrating escalation or addiction.

Q3. What influences, motivations, and questions led you into experimental gaming/economics? Why experiment? What were your thoughts on this, as you now recall?

Q4. What were the origins of your collaboration with Fouraker and Siegel? When did it begin, before or after their first work in bargaining (published in 1960)? What were your impressions of Siegel, and how he got into experimental research on bargaining/oligopoly? How do you see your influence on them; their influence on you?

A. Concerning actual experimentation in economics, this did not occur to me until 1955 (1956?) when Martin Beckman and I met Sidney Siegel around a campfire in Tuolomne Meadows in Yosemite. Beckman and I (instead of singing "Clementine") were discussing utility theory—when a shadowy figure near us started to eavesdrop and then could not contain himself and joined in the conversation. From there I learned about Siegel's two-light experiment with rewards—as I knew Estes's learning work I was immediately fascinated. I immediately said to Sid that Mayberry, Nash, and I had written an article on duopoly (Cournot and Cooperative) and did he think that this might be worthwhile. He said yes, but that an economist at Penn State—Larry Fouraker—had already talked with him about bilateral monopoly and that he was committed to finish that work before looking at duopoly.[7] We agreed to join forces in two or three years—after their work on bilateral monopoly was finished and if I was in a position to collaborate—and had planned a fairly lengthy collaboration.

The next influence came in a different direction—Dick Bellman was a good friend of mine and he told me about the commercial venture he had been hooked into by the American Management Association—the first major computer business game (Bellman, et al. 1957). In the meantime, in September 1956 I joined General Electric and fairly soon met George Feeney. It seemed to both of us that, first of all, business games were a great teaching tool—but also there was no reason why they should not have enough structure that one could also analyze them experimentally.

While at GE, I became an adjunct research professor at Penn State and went out to work in '59, '60, '61 with Siegel, Fouraker, and a bright graduate student, Harnett. (Saturday, 25 June

7. In a letter to me [Vernon Smith] dated 20 September 1990, Lawrence Fouraker states: "My interest in the experimental testing of economic theory started with a master's candidate, Dave DiFedo, who wanted to test my bilateral monopoly model against Fellner's. I knew of Siegel's experiments on choice theory and suggested we talk with him. Sid liked the idea and had several research assistants and considerable experience in organizing and conducting experiments. Most of this is covered in *Decision and Choice* by Messick and Brayfield (1964). I knew of Chamberlin's theoretical work but nothing of his experiments."

1960, Sid and I ran experiments on Edgeworth cycles.) Sid and I had agreed that after the experiments with Fouraker on duopoly and triopoly we would like to computerize like crazy—I felt that the business games had pointed the way, especially the one that Feeney had built and we were using at GE. Sid died before we even finished the work we were doing.

While I was at GE, I was also in touch with John Kennedy, the psychologist at Princeton, and was essentially a shadow thesis committee for David Stern whom we supervised in doing an excellent thesis on oligopoly experiments in 1960. His thesis is available and still worth reading. (He then became an assistant professor of economics at UCLA.)

Kennedy stressed to me the dangers of getting results by appropriate briefings. He had a great deal to say concerning the sociopsychological aspects of artifacts in gaming.

On Wednesday, 20 January 1960, George Feeney and I ran an experiment at the Stanford Research Institute (SRI) with a four-team business game run by about 20 players and around 150 individuals trading shares in a stock market run by four stockbrokers. The results were published in 1970 (Shubik 1970).

I cannot remember precisely when I met Hoggatt—but my best guess is around 1958—we talked mainly simulation to start with, then the idea of a gaming lab.

In September 1960, I was a visiting professor at Yale for the year. On 5 October 1960, I talked with Masuo Toda of Tokyo on his 3 × 3 matrix games results. February 1961, I did 2 × 2 matrix game experiments with limited information (Shubik 1962). On 12–15 April 1961 I stayed with Sid Siegel, finishing the design of quantity and price-quantity oligopoly. I saw Larry Fouraker on the 13th. We ran 12–14 duopoly pairs under incomplete information.

At Yale in 1960, I gave a small seminar on oligopoly theory and gaming—three students, Jim Friedman, Ted Tarson, and Tappen Roy—all three plus myself were employed in building my oligopoly game which in essence was the step that Sid and I had eventually intended to take. It forms the basis for my book with Levitan (Shubik and Levitan 1980).

Friedman was also (in my opinion) influenced by George Feeney—he did his thesis on experimental games.

I went to IBM in October 1961 and started to work with Dick Levitan to move my game from an IBM 650 to a bigger, better machine—also to add any new features to it. (Our first game used a template as we had no way to print out the format. We could only print numbers.) Levitan did his thesis with Dorfmann and me on a quadratic programming method for allocating demand among oligopolistic firms with product differentiation.

I also persuaded C. West Churchman to have Systems Development Corporation (SDC) sponsor in 1962–63 an enormous set of experiments with the oligopoly game for 1, 2, 3, 4, 7, and 10 competitors—in all I guess of the order of 100 experiments—Eisenberg at SDC was involved and later Gerritt Wolf at Yale. The results were partially published in 1972 (Shubik, Wolf, and Eisenberg 1972).

The people who basically really influenced me (and possibly vice-versa) were the group at Princeton—playable games to illustrate paradoxes. Kennedy, Bill Estes, Sid Siegel—experimental psychology counts and can be used in economics. Dick Bellman, George Feeney, Austin Hoggatt—computer games are the way to go.

Later I was obviously stimulated and influenced by Selten, Rapoport, Plott, Roth, Smith, Friedman, Sauermann, Guth.

My original goals 1950–55 were games for illustration, then after 1955 the idea of checking Cournot and Bertrand became clear and in particular the studying of limiting processes—how competition changes under 1, 2, 3, 4, . . . competitors. Then by 1962 the idea that the eventual goal should be the fully computerized game hopefully with multidisciplinary experiments. This was in part realized with work with Gerritt Wolf after Sid died. My current interest is to find or build a group to collaborate with me on the running of my money models from the theory of money and financial institutions.

Q5. When did you first learn of the work in Germany on experimental economics (Sauermann and Selten 1959, English translation 1960)? As you see it, was this work an independent development, or were there cross-country influences of which you were aware?

A. I probably heard of the work in Germany in the mid-60s. It was probably around then that I stumbled on Chamberlin's article.

Herbert Simon

Herbert Simon was one of the first scholars in the 1950s to develop an interest in experimental games/economics. He attended the 1952 Santa Monica conference on "The Design of Experiments in Decision Processes," which led to Thrall et al., *Decision Processes* (1954). His *Psychometrika* paper (Simon 1956a) directly cites the stimulus of the 1952 conference. His influence on early experimental games/economics was so important that his early impressions are of particular significance.

Q1. Generally, what are your impressions of the origin of the interest in experimental games/economics?

A. I am not sure that I can be very helpful regarding the early history of game theory, and especially experimental gaming, something I was never involved in. I first became interested in game theory (before the publication of von Neumann and Morgenstern) in the 1930s because of a concern with the outguessing problem in social prediction. Hence I noted the von Neumann and Morgenstern book immediately on its appearance and wrote the very first review of it, published in the *American Journal of Sociology* in 1945. I was then participating in Cowles Commission seminars and shortly thereafter became a consultant to the RAND Corporation, so I came to know von Neumann, participated in the 1952 summer seminar, etc.

However, I was mostly interested in *n*-person games, and was profoundly dissatisfied with the concept of "solution" in von Neumann and Morgenstern—it seemed to me to confirm the complexity of the problem rather than solve it. Hence, game theory played a relatively restricted role in my own thinking (in spite of the one paper you mention). Economists were, of course, immediately interested in game theory as a possible answer to the intractable problems of oligopoly, but many of them were quickly disillusioned in the prospects of getting answers from the theory. I think there was a decline and a more recent revival in game theory that you can probably verify from publication rates etc.

I do not think that the impetus for experimentation within a game-theoretical framework initially came from economists, but rather from psychologists (particularly those who had begun to build mathematical learning theory), statisticians, and interdisciplinary types close to cybernetics and management science (see the list under Q2 below). Siegel, who I always supposed was the key member of the Siegel/Fouraker team, was a psychologist. Experiments, of course, came naturally to psychologists—that is the way they learned to do science. (The references in the Siegel/Fouraker book bear out this interpretation.)

Q2. Who were the principal contributors? Chamberlin (1948) was apparently first in economics. Others seem to include Merrill Flood, Anatol Rapoport (1950s) in addition to Siegel, Fouraker and Siegel (1960, 1963), Austin Hoggatt, Sauermann and Selten, M. Shubik, and James Friedman. But what others?

A. These comments also bear on your third question. There was a lot of interest in the postwar "cybernetic" community in experimentation (not necessarily in a game-theoretic framework), but it came largely from the social psychologists resident in that community—including Bush, Bavelas, Guetzkow, and myself, the Systems Research Lab at RAND (Kennedy, Chapman, and Newell), and others of that sort. Shubik did mainly theory, although he was involved in empirical modeling (not experimentation) at GE. My conclusion is that most of the economists and statisticians who became involved did so by brushing up against psychologists at RAND and similar places, and learning something about experimentation from them. But, it never really caught on among economists until you and then Charlie Plott (I have always supposed, independently of these earlier developments) got into the act. But you know that part of the history far better than I.

Q3. What impressions do you have of the intellectual influences, motivations, and questions that led to the 1952 Santa Monica conference and subsequent developments? Why was there interest in experiments only a few years after the publication of *The Theory of Games and Economic Behavior*? Such has hardly been a natural follow-on interest for economic theorists generally, nor among game or economic theorists today (with notable recent exceptions)?

A. As I have already suggested, I believe that the 1952 Santa Monica conference came out of the general stir about the whole range of things that was then sometimes put under the label of cybernetics. RAND was at the center of that stir, and just about everyone involved had close connections with the RAND group and/or the Cowles Commission. This was a response not only to von Neumann and Morgenstern (which was itself a response to these developments), but to the whole postwar interest in the applications of mathematics to human affairs—which encompassed computers, servomechanism theory, information theory, mathematical economics, mathematical learning theory, game theory, brain architecture, robots, and operations research (I am sure I have omitted some items). To the extent that some of the people interested in these matters had backgrounds in various areas of empirical science, they brought empirical techniques, including experimentation, into the picture.

As I reread these more or less random comments, I see I have omitted Princeton as one of the key centers of these developments. Naturally, many of the key young people at the 1952 meeting were Princeton mathematicians, including Shubik, Nash, Shapley, etc. My best guess is that they picked up the idea of doing experiments from having played games with each other while at Princeton. They were not very sophisticated in their experimental methodology, but of course were very sharp on the theory side.

For another view of the scene, you might want to read the Historical Addendum, p. 873 ff. of Newell and Simon, *Human Problem Solving*.

Q4. What were your impressions of Siegel and how he got into experimental research in bargaining/oligopoly?

A. I did not know Siegel well, but he was always highly spoken of as an extremely careful experimenter who also had an excellent command of the underlying theory.

James Friedman

James Friedman was one of the early and influential contributors to experimental games/economics. He was directly influenced (as I was) by the work of Siegel and Fouraker and by that of Shubik, and although his intellectual interests have concentrated in game theory, he has always been a strong supporter of experimental methods in economics.

Q1. Generally what are your impressions of the origin of the interest in experimental games/economics? In particular, what and who influenced your entry into experimental research? When did your work begin?

A. I went to Yale in 1959 especially interested in economic theory, but during my first two years the theorists were mostly absent (Koopmans on leave both years, Debreu and Marschak on leave year 1, then accepting new positions). Martin Shubik was a 1960–61 academic-year visitor who returned permanently in 1963. As a second-year student I, with two fellow students, took a two-semester course from Martin taught from *Strategy and Market Structure* [1959] and *Games and Decisions* [Luce and Raiffa 1957]. Oligopoly theory, part of Fellner's microtheory core course the year before, had particularly interested me and that interest was stimulated further in Martin's course.

During the year with Martin, he was working to design a business game that would be more analytically tractable than the typical business games of the time. He enticed me as an (unpaid and delighted) research assistant to help. When the school year was nearly over he asked about my dissertation plans, which were vague, and suggested I do experimental research. The thought had never crossed my mind, but instantly the idea was appealing. At that point I'd never heard of Siegel, Fouraker, Smith, or anyone who had done an economics experiment except for Mosteller and Nogee whose utility experiment had been part of Willy Fellner's course. Soon after, Tjalling Koopmans brought the work of Siegel and Fouraker to my attention and lent me two fat working papers that were the manuscript of *Bargaining Behavior* [Fouraker and Siegel 1963]. *Bargaining Behavior* was an immensely important tutorial for me on how to run and document experiments and also provided a benchmark from which I figured out my experimental intentions and design. Martin was at IBM then, but we kept in touch. George Feeney had replaced Martin as a one-year visitor. Martin urged me to just replicate the oligopoly experiments in *Bargaining Behavior*, but I was unwilling to follow his scientifically sound advice because my work would then be insufficiently "original."

During the rest of the year I had close contact with George Feeney, talked occasionally to Willy, who was my chairman, and ran my experiments in March and April of 1962. Both helped me, and I learned a great deal about experimental economics from George.

My next experimental work was the duopoly with communication experiment (a cheap talk experiment) that appeared in *Econometrica* (Friedman 1967) and in a companion piece a couple of years later in a volume edited by Heinz Sauermann. While I was at Berkeley for a year in 1966–67 Martin urged me to contact Austin Hoggatt. Auggie and I embarked on the work that was finally published in your series (Friedman and Hoggatt 1980). The collaboration with Auggie was smooth and delightful from start to finish. We certainly disagreed any number of times, but we were always wanting to listen to one another to figure where the root of disagreement lay and to figure out what we should do.

Q2. When did you first learn of the work in Germany?

A. When I first met Reinhard Selten, which was probably late summer of 1967. Reinhard spent the 1967–68 academic year at Berkeley where his long collaboration with John Harsanyi began. Sometime during that year, probably at the beginning of it, he was in New Haven for a few days. He solicited the sequel to my 1967 *Econometrica* paper for one of the Sauermann volumes and, while I don't specifically recall it, he must have told me about the experimental work going on in Germany at that time.

Q3. Why do experiments?

A. For me the great motivation and prime interest lies in testing the predictions of economic theory. When you test a theoretical prediction on naturally occurring data you do not find out whether the prediction will hold under the assumptions of the model from which the prediction comes. The reasons are that naturally occurring data inevitably arise in circumstances that don't satisfy the conditions of our models and the correct conditions are imperfectly known. What you find may be valuable in making policy, but the foundations are shaky when you lack good knowledge of the underlying model that generates the naturally occurring data. You don't have a clear picture of why the given prediction works (or fails to work, as the case may be). To build theory from solid foundations and to build a strong structure, I believe that testing predictions in conditions that satisfy the assumptions under which the predictions are made is the first step in gauging the validity of such predictions. This sort of testing does not imply that the same predictions will obtain on natural data; however, it is an essential step in seeing whether a theory is valid on its own terms.

Reinhard Selten

Reinhard Selten was one of the very first, and certainly influential, contributors to experimental games/economics. He clearly merits credit, along with his teacher Heinz Sauermann, for founding the German movement in experimental economics. His responses are particularly valuable in helping us to see why experimental economics in Germany and the United States were coincident developments in time.

Q1. Generally, what are your impressions of the origin of the interest in experimental games/economics? In particular, what and who influenced your entry into experimental research? When did your work begin?

A. I was influenced by two publications. One was the paper by Kalish, Milnor, Nash, and Nering on their pioneering characteristic function experiments (Thrall et al. 1954). The other was a little book published by the A.M.A. (American Management Association) on a computerized business game. The first publication encouraged me to make experiments and the second one suggested the field of oligopoly. However, there were some further influences.

During the time I studied mathematics I also attended many lectures in psychology. Like other psychology students I had to participate as a subject in psychological experiments and also help to perform them. My psychology teacher was the gestalt psychologist Rausch who taught at Frankfurt at this time.

My first experimental paper, "Ein Oligopolexperiment," *Zeitschrift für die gesamte Staatswissenschaft* 1959 (Sauermann and Selten 1960), was also the first of all my papers. It was written with Heinz Sauermann, my economics teacher. I was lucky to succeed not only in convincing him of the appropriateness of experimental work in economics but also in getting him involved in the performance of experiments. As you probably know he later organized a series of conferences on experimental economics and edited the book series *Beiträge zur experimentellen Wirtschaftsforschung* (Sauermann 1967, 1970, 1972).

Q2. Who were the principal contributors?

A. I continued to do experimental research in the sixties. The experimental research I did in the late fifties and the sixties is published in the first two volumes of the *Beiträge zur Wirtschaftsforschung*. In these volumes—not only the first two—you will also find papers by other assistants of Professor Sauermann who, under my guidance, did experiments: Becker, Berg, Haselbarth, Tietz, and others. Among the early contributors I would also include Thurston (1931) who tried to measure indifference curves.

Q3. What accounts for the "sudden" interest in experiments around midcentury? The Santa Monica conference in 1952, "The Design of Experiments in Decision Processes," led to Thrall et al., *Decision Processes* (1954). That is not long after *The Theory of Games and Economic Behavior*. That early interest in experiment by game/economic theorists hardly typifies the tendencies among theorists today.

A. My impression about the reason why early experimentation started in the late fifties can be summarized as follows. Oligopoly theory had created a profusion of different approaches. After the rise of game theory one might have hoped for some improvement in this respect but cooperative game theory also soon offered a multitude of conflicting solution concepts. It became increasingly clear that this kind of rational theory would not be likely to produce a satisfactory solution of the oligopoly problem or the coalition formation problem. In this situation, experimentation was an idea that suggested itself especially to people who had some contact with experimental psychology. Fouraker was lucky to meet the gifted psychologist Sidney Siegel.

I have already told you about my personal motivation to begin to do experiments. At first I hoped to find an experimental theory of oligopoly. A short time after I had begun to do experiments, I read Simon's paper on bounded rationality in the book *Models of Man* [1957]. From then on, my own interest in experimental economics became more and more connected to the goal of developing a theory of bounded rationality. I am still working toward this goal but I fear that I shall not succeed in my lifetime. Up to now we have only bits and pieces of such a theory. I see the main task of experimental economics to be in this direction.

The 1952 Santa Monica Seminar[8]

In the summer of 1952 the Ford Foundation sponsored an eight-week seminar in Santa Monica on the topic "The Design of Experiments in Decision Processes." The seminar resulted from a proposal by an interdisciplinary group at the University of Michigan, which had grown out of a regular seminar initiated in the fall of 1950 under the joint sponsorship of the departments of economics, mathematics, philosophy, psychology, and sociology. The Santa Monica seminar was attended by thirty-seven participants. The choice of location made it possible for a number of people to attend under RAND Corporation sponsorship. The result was an impressive interorganizational and interdisciplinary venture: among the participants, seven were sponsored by Ford; twenty by RAND; two by ONR (the Office of Naval Research); one by the Cowles Commission; and seven had joint sponsorship by two of these four organizations.[9] The roster of already- or soon-to-become distinguished scholars (including three Nobels) who either participated or who were among the contributors to Thrall, Coombs, and Davis (1954; hereafter TCD) included such well-known luminaries as the psychologists R. Bush, C. Coombs, W. Estes, L. Festinger, and H. Simon; the mathematicians S. Karlin, J. Nash, L. Shapley, and J. von Neumann; the economists G. Debreu, C. Hildreth, T. Koopmans, J. Marschak, O. Morgenstern, and R. Radner; and statistician/decision theorists such as F. Mosteller and H. Raiffa.[10] Experimental economists unfamiliar with

8. Much of the source material for this section is derived from Thrall, Coombs, and Davis 1954.

9. An interesting paper by Mirowski (1990) on the early history of game theory develops the social constructionist thesis that "the connections between the military and game theory were numerous and pervasive in the first two decades of its existence, extending into the very mathematics itself" (Mirowski 1990, 1). In the Santa Monica seminar, 27 of the 37 listed participants were sponsored in full or in part by either RAND or ONR. However, none of the papers presented and none of those published in Thrall, Coombs, and Davis (1954) dealt, or was concerned, with military applications of game theory or decision theory. In this case military support appears to have been for basic research unconnected with military applications.

10. Some who presented papers (for example, J. von Neumann) were not listed as participants.

TCD will not fail to be surprised that all this intellectual power congregated in response to the theme "the design of experiments in decision processes." What the supporters of the Santa Monica seminar hoped was to provide a presentation-discussion forum that would stimulate empirical research and further theory development; quantitatively, its success in stimulating experiment must be judged much more modest than in furthering theory.[11] The seminar's integrating theme was not experimental design, but the use of mathematics in the social sciences. Only five of nineteen chapters are concerned primarily with reporting the results of experiments. But, I shall argue below, the influence on subsequent experimentation, and upon fundamental concepts, through H. Simon and S. Siegel, was very significant and lasting. Also Marschak's subsequent contributions to experimental economics must be credited in part to the seminar's influence. Finally, from the statement by Reinhard Selten above, the experimental contribution by Kalish, Milnor, Nash, and Nering (in TCD) was a primary influence on his development, and, as is universally recognized, Selten was the catalyst in founding the German experimental economics program.

I propose to focus this discussion on the three papers dealing with the "Bernoulli Choice" experiment, as I shall call it, or "probability learning" under which it has been studied in psychology (Estes, chapter 9, and Flood, chapters 10 and 18, in TCD). This experiment generated considerable discussion and controversy at the Santa Monica seminar and directly stimulated new research on the part of some of the seminar participants; these in turn influenced important research studies by Sidney Siegel, which I shall treat in the next section.

On each of many trials the subject chooses between two (or more) alternatives A_1 and A_2, associated with events E_1 and E_2. If A_i is chosen and E_i occurs, the subject is "rewarded" (Simon 1956a, 267) with some

11. The Santa Monica conference may have had an indirect influence on me through Jacob Marschak, who visited Harvard in 1953 to give a seminar. He also held office hours for the graduate students, and I took advantage of this opportunity. In this session he talked about decision theory and behavior and pointed out that there was a simple method for getting a buyer to reveal his maximum willingness-to-pay for an item. A seller writes his selling price on a card and places it face down on the table. The buyer then announces her buying price with the understanding that she makes a purchase at the seller's price if and only if the buying price exceeds the selling price. What I learned from the session and this example was how operational and doable economic theory could be. Eventually, I did experiments with the "Marschak" auction (Coppinger, Smith, and Titus 1980), which of course anticipated Vickery's analysis of the second price auction by eight years, and was, to my knowledge, the first brush with the important concept of incentive compatibility and of a dominant strategy.

probability π_i. "The subject is given no information about the conditions of the experiment except that some one of the E_j will follow the signal S (to make a choice) on each trial" (Estes, in TCD, 128). In its simplest structural form, the Estes model (also see Simon 1956a, and Bush and Mosteller in TCD) describes the subject's choice behavior as a stochastic process, with (for $n = 2$ events)

$$p_i(t+1) = \pi_i p_i(t) + (1 - \pi_j)(1 - p_i(t)) \tag{1}$$

where $p_i(t)$ is the probability that the subject will choose A_i on trial t.[12] Asymptotically, in a convergent "learning" equilibrium, we have $p_i(t+1) = p_i(t) = p_i^*$, and from (1), if $\pi_1 + \pi_2 = 1$, then (Estes, TCD, 130) we get the asymptotic prediction

$$p_i^* = \pi_i \tag{2}$$

At the Santa Monica seminar this model was the center of some controversy, which became of considerable historical significance. As reported by Flood (also see Simon 1956a, 267–68), "After Estes had presented his paper, various game-theorists in the audience argued (incorrectly) that this observed behavior was surprisingly 'irrational,' since the organism's proper strategy is clearly pure, rather than mixed, and so the organism should learn eventually to choose only the alternative providing the more frequent reward" (Flood in TCD, 287–88). Flood's description of the theorist's argument as "incorrect" was based on two objections. First, in applying game theory it is unclear which payoff utilities are to be used. For example, one of Flood's subjects remarked (in an experiment with $\pi_1 = 0.90$, $\pi_2 = 0.10$) that the only way to get a perfect score would be to choose E_1 90 percent of the time, E_2 10 percent of the time, and to be "lucky on each play" (Flood in TCD, 288). Second, game theory cannot be applied because it cannot be supposed that the subject believes that the events constitute a stationary stochastic process; in particular the process might be thought to contain predictable patterns. Ultimately, this last argument would become moot, in the sense that it was subsequently shown (see, for example, Siegel 1961, Tversky and Edwards 1966) that the "probability matching" behavior reported by Estes was replicated even where care was taken to inform the subjects

12. In addition to being reward probabilities, π_1 and π_2 are also the conditional asymptotic probabilities that the subject will *persist* after having just chosen A_1 or A_2; the corresponding conditional probabilities of *shifting* are $1 - \pi_1$ and $1 - \pi_2$.

that the process they were observing was the result of independent trials with fixed probabilities of occurrence.

Davis, in his "Introduction to 'Decision Processes,' " discusses the division of points of view—particularly empirical versus theoretical—that were evident at the seminar (TCD, 3–5, 10–11, 16–17, passim). Concerning Flood's comments on the game theorists, Davis writes parenthetically that "(the editors feel that his discussion of the remarks of 'various game theorists in the audience' may be misleading. It is our recollection that in his initial report Estes left an impression with the audience that his subjects had reason to believe that the process was random, and we further recall that no one failed to preface remarks about 'rationality' or 'proper strategies' with a more or less explicit assumption about payoff utilities, such as '*If* they wanted to maximize scores . . .')" (16–17).

The importance of this interchange is that Herbert Simon, taking his cue from Flood, used this example to develop the essential distinction between objective and subjective rationality which underlay his contrast between substantive and procedural rationality, and his notion of bounded rationality. First, he noted that "When this (Estes's) experimental situation was described to a number of game theorists at the Santa Monica conference, they pointed out that a '*rational*' individual would first estimate, by experimenting, which of the two alternatives had the greatest probability of reward, and would subsequently always select that alternative" (Simon 1956a, 267–68). He then used a derivation to show that the Estes equilibrium rule (2) was implied for an agent desiring to minimax regret à la Savage. More significantly, he draws some conclusions from this exercise which, I believe, provide one of his best articulations of the ideas for which he is most celebrated in economics and which directly influenced Sidney Siegel.

> We need not try to decide whether the subjects who behave in conformity with the predictions of Estes' theory are minimaxing regret, or whether they are simply behaving in the adaptive fashion implied by the usual learning mechanisms. Most economists and statisticians would be tempted to accept the former interpretation, most psychologists the latter. It is not immediately obvious what source, other than introspection, would provide evidence for deciding the issue. . . . Perhaps the most useful lesson to be learned from the derivation is the necessity for careful distinctions between *subjective* rationality (i.e.,

behavior that is rational, given the perceptual and evaluation premises of the subject), and *objective* rationality (behavior that is rational as viewed by the experimenter). Because this distinction has seldom been made explicitly by economists and statisticians in their formulations of the problem of rational choice, considerable caution must be exercised in employing those formulations in the explanation of observed behavior.

To the experimenter who knows that the rewards attached to the two behaviors A_1 and A_2 are random, with constant probabilities, it appears unreasonable that the subject should not learn to behave in such a way as to maximize this expected gain—always to choose A_1 (if $\pi_1 > \pi_2$). To the subject, who perceives the situation as one in which the probabilities may change, and who is more intent on outwitting the experimenter (or nature) than on maximizing expected gain, rationality is something quite different. If rationality is to have any meaning independent of the perceptions of the subject we must distinguish between the rationality of the perceptions themselves (i.e., whether or not the situation as perceived is the real situation) and the rationality of the choice, given the perceptions.

If we accept the proposition that organismic behavior may be subjectively rational but is unlikely to be objectively rational in a complex world then the postulate of rationality loses much of its power for predicting behavior. To predict how economic man will behave we need to know not only that he is rational, but also how he perceives the world—what alternatives he sees, and what consequences he attached to them . . . [Simon 1955]. We should not jump to the conclusion, however, that we can therefore get along without the concept of rationality. (Simon 1956a, 271–72)

A question not part of the public record (in TCD) at the Santa Monica conference is whether behavior is influenced by the use or absence of explicit monetary rewards for a prediction of the events E_1 and E_2. Simon, in discussing the Estes environment, writes, "Each alternative is rewarded on a certain per cent of the trials" (Simon 1956a, 267). But it is unclear what "rewarded" means. Bush, Mosteller, and Thompson state that "the experimenter may reward one response and punish another. . . . A pellet of dog chow may be rewarding to a hungry rat, but a nod of approval may be equally effective for an 'ego-involved' human subject" (TCD, 99). Specifically in his experiments, Estes reports that "the subject is instructed to do his best to make a good score (in terms of correct predictions)" (TCD, 128). Similarly, in both of the experiments reported by Flood that were motivated by the Estes paper and its subsequent discussion, the subjects were instructed thus: "Your object is to get as many wins as you can" (Flood in TCD, 297, 298). None of these (human subject) experiments used any explicit reward of the kind usually thought of as utilitarian.

Siegel, Siegel and Fouraker, and the Development of Experimental Economics

The Bernoulli Trials Experiment and Subjective Rationality

Siegel (1959, 303–5; 1961) cites Simon in the quotation above—on distinguishing subjective and objective rationality—as the motivation for his work in the two-choice uncertain outcome experiment. Siegel, like Flood and Simon, reinterprets what it means to be subjectively rational in this task, but takes a utility-maximizing approach. He postulates that in this experiment, the only "payoff" is the satisfaction of seeing one's prediction confirmed or the dissatisfaction of seeing it not confirmed. Furthermore, there is the kinesthetic monotony of pressing the same button (corresponding to the more frequent event) for hundreds of trials, and the cognitive boredom of "thinking left" trial after trial after trial. In effect Siegel suggests that in the binary choice task psychologists have wrought a boring experiment without monetary rewards; in such circumstances a mixed strategy of choosing between the two events may serve to maximize subjective satisfaction. What is needed is to analyze the problem from the point of view of the bored subject. To this end, at the first level of analysis, Siegel postulates two additive components of subjective expected utility: the utility of a correct response, $a[p\pi + (1-p)(1-\pi)]$, where a is the marginal utility of a correct prediction, and the utility of variability which is assumed to take the particular form $bp(1-p)$, where b is the marginal utility of variability. People seek variability to relieve monotony, and this particular form of the utility of variability has the required property that it is maximal where $p = 0.5$; *that is*, when relief from monotony is greatest. Total utility is thus

$$U(p) = a[(1-\pi) + p(2\pi - 1)] + bp(1-p), \quad \pi > 1/2, \tag{3}$$

Table 1 Test 1: Mean and Variance by Reward Condition on Final Block of 20 Trials for $\pi = 0.70$

	Reward Condition		
	No Payoff	Payoff	Payoff−Loss
Number of Observations	12	12	12
$\overline{p}$	0.70	0.77	0.93
$S^2 = 1/11[\Sigma_i(\hat{p}_i - \overline{p})]$	0.00477	0.00385	0.00335

which is concave in p. The strategy p^* that maximizes (3) yields

$$p^* = \frac{a(2\pi - 1)}{2b} + 1/2 = \alpha(\pi - 1/2) + 1/2, \tag{4}$$

where $\alpha = a/b$. From (4) $p^* = \pi$ if any only if $\alpha = 1$; i.e., when the marginal rate of substitution of variability for a correct prediction is unity. More generally (allowing for a boundary maximum):

$$p^* = \begin{cases} 1, & \text{if } \alpha \geq \frac{1}{2\pi - 1} \\ \alpha(\pi - 1/2) + 1/2, & \text{if } \alpha < \frac{1}{2\pi - 1}. \end{cases} \tag{5}$$

According to (4) an increase in a and thus α, for given b, will increase p^*. The most obvious way to increase a is to introduce monetary rewards. This was accomplished by randomizing 36 subjects into three groups of 12: in one group each subject received no salient reward, only a fixed payment for his or her participation; in the second group each subject received $0.05 for each correct prediction; each subject in the third group received $0.05 for a correct prediction and lost $0.05 for an incorrect prediction. These three treatments correspond to increasing values of a, and the three groups are predicted to exhibit the asymptotic choice behavior: p^* (no payoff) $< p^*$ (payoff) $< p^*$ (payoff $-$ loss). Table 1 lists the mean observed values, $\overline{p}$, and the variance (which we interpret as mean square decision error) on the final trial block for each treatment. Note that the observed values of $\overline{p}$ increase, and the mean square "error" decreases, with increases in the motivation to predict correctly.

Based on this first experimental test, Siegel concludes: "Thus the data lend strong support to the contention from decision-making theory that a person behaves in this situation as if he were attempting to maximize expected utility" (1961, 772).

Next Siegel notes (772) that the first test, while supporting the qualitative predictions of the model, does not provide a test of its particular features. He accomplished this by two additional experiments. In the first he randomizes 40 subjects into two groups. The first is run under the payoff − loss condition, with $\pi = 0.75$, and the observed choice proportions are used to compute the parameter α in (4). From this value of α he predicts the asymptotic value of p^* in the second group, tested with $\pi = 0.65$, and finds that the latter deviate insignificantly from the prediction. In the second experiment he tests directly his hypothesis about the utility of variability by introducing a treatment designed to decrease b in (3), while holding a constant. This test was "based on the notion that the utility of variability can be reduced by giving the subject an opportunity to make the same prediction, for example 'left light' by different responses" (774). This was accomplished by using a swivel chair to randomly cause the subject to face either the light board offering two prediction buttons, or a mirror reflecting the light board, and also offering two prediction buttons. By this procedure Siegel sought to introduce cognitive variability (via the mirror) and kinesthetic variability (via the swivel chair)—both designed to reduce the monotony of the task. The results of the second experiment showed that the procedure increased the asymptotic value of $\overline{p}$, in accordance with the model when b is increased. Hence, by modifying rational theory along lines that he acknowledged were stimulated by Simon (1956a) and by the subsequent skillful use of experimental design, one could arrive at a rather different conclusion than the following: "The (Estes) learning theories appear to account for the observed behavior rather better than do the theories of rational behavior. . . . However adaptive the behavior of organisms in learning and choice situations, this adaptiveness falls far short of the ideal of 'maximizing' postulated in economic theory. Evidently, organisms adapt well enough to 'satisfice'; they do not, in general, 'optimize' " (1956b, 261). What Siegel showed was that people did not maximize the expected number of correct predictions in the Bernoulli trials experiment because it was not in their interest to do so. This failure to "maximize" was thus actually the exception that proved the (subjective utility maximizing) rule.

A subsequent interpretation of Siegel's research was that it instructs us on the importance of "other things" in the utility function besides monetary reward (Smith 1976a), and in particular that all decisions have implicit decision costs (Smith and Walker 1990), which are not part of the concept of objective rationality as ordinarily modeled. Rationality must be interpreted from the perspective of the decision-maker in terms of his/her actual experience of the decision process and its consequences. This perspective blurs the sharp distinction usually drawn between normative and descriptive models; between substantive or objective rationality, and procedural or subjective rationality; and it provides a better predictive model of rationality as it actually might be experienced by the individual. Siegel's work is exemplary in showing what psychology can contribute to economics by modifying decision theory in the light of rigorous experiment. In contrast, many psychologists today who study decision making emphasize the predictive failures of "rational" theory; they tend not to ask with Siegel whether it might be less than rational to be "rational."

Payoff Opportunity Cost and Deviations from Equilibrium

Siegel's earliest concerns for providing monetary motivation of experimental subjects carried over into his work with Fouraker in bilateral bargaining and oligopoly. In their first collaboration the results of five experimental sessions in cooperative (they called it "equal strength") bargaining were reported (Siegel and Fouraker 1960; hereafter SF). In these experiments subject dyads were allowed to freely bargain through a structured alternating series of price-quantity messages until a given message by one party is accepted by the other party, yielding a contract. This is a cooperative game in approximately the sense defined in Luce and Raiffa (1957, 89), wherein the participants are permitted to engage in preplay communication and make binding commitments. The theoretical prediction—an equal split of the joint maximum (Pareto-optimal) surplus—was supported by the results of experiment 1, but with considerable dispersion among the bargaining pairs (SF, 25–27). This experiment was replicated twice (experiments 4 and 5 in SF, 35–40) with a treatment using a modified payoff table that substantially increased the payoff difference between the predicted equilibrium and one-unit quantity derivations therefrom. This opportunity cost (payoff difference) treatment dramatically decreased the number of bargaining contracts that failed to be Pareto-optimal: in experiment 1, 7 of 11 contracts failed to maximize joint payoff; in experiments 4 and 5 pooled, only 1 of 22 contracts failed to maximize the joint return.

These sensitivity exercises were continued in Fouraker and Siegel

(1963; hereafter FS) in the context of two noncooperative games: bilateral bargaining in which the seller moves first, choosing price, and the buyer moves second, choosing quantity (posted offer); quantity adjuster (Cournot) duopoly and triopoly in which sellers choose quantities and demand is simulated to be fully revealing. In three of the bargaining experiments, payoffs were tripled relative to those in the baseline experiments (FS, 227, 233, 239). In two of the Cournot experiments, bonuses were paid to the top three profit makers (FS, 164–65, 304–6). None of the augmented payoff treatments changes the central tendency of the results to support the Nash equilibrium hypothesis. However, in three of the five tests the error variance was reduced, and in the remaining two the failure of the error variance to narrow was the result of a single outlier observation (Smith and Walker 1990, table 1).

These early concerns for the effect of reward level and opportunity cost on observed decision outcomes have influenced the course of experimental economics down to the present. Since Siegel, Siegel and Fouraker, and Fouraker and Siegel, there are numerous examples of studies in which experimentalists have examined the robustness of results to payoff treatments (see Smith and Walker 1990 for a summary and general model of decision cost in experiments). In spite of this heritage one still sees it claimed in the psychology literature that monetary rewards make "little" difference (Tversky and Kahneman 1986, 90; Thaler 1986, 96). Also, in spite of this long history, we have witnessed the rediscovery of the principle that payoff opportunity cost might be important in experimental economics, and the claim made, without empirical demonstration, that payoff opportunity cost is generally too small (Harrison 1989)—a hypothesis that can only be evaluated empirically.

In Smith (1982; also see 1976a) it is argued that the implications of Siegel 1961 are the following: a sufficient condition for a valid experimental test of propositions from neoclassical rational theory is that monetary rewards in an experiment dominate nonmonetary arguments in the subject's utility function of experimental outcomes; this condition is sufficient, not necessary, because dominance may not be achievable at any levels of payoff, in which case the research program must necessarily be directed to modifying the traditional theory in the light of experimental evidence on the effect of varying payoff levels. Finally, it should be noted that the problem of nonmonetary motivations (decision cost) is identical to the problem of low opportunity cost of deviations from the rational optimum. Simply put, this is because traditional theory predicts that agents will go to their unique utility maxima however relatively flat is the utility hill. If flatness matters it is because there exist nontraditional utilitarian elements, and the agent faces trade-offs which cannot be ignored in studying agent behavior. Such trade-offs need to be modeled and to be studied experimentally.

Complete versus Incomplete Payoff Information in Markets

Cooperative Bargaining. In their study of "cooperative" bargaining, SF draw a distinction between complete and incomplete (private) information that continued into the FS study of noncooperative bargaining and oligopoly. In two-person bargaining, under complete-complete (hereafter complete) information each subject knows his own payoff and that of his opponent for every possible contract. Furthermore, each knows that his rival has the same information. Effectively, this is equivalent to what game theorists today call common knowledge, where the philosopher Lewis (1969) is cited as the source of this inspiration (compare Aumann 1987). Under incomplete-incomplete (hereafter incomplete) information each individual knows his own payoff for every possible contract but does not know his rival's payoffs; neither knows how much information the other possesses.[13] Complete-incomplete information is where one individual knows his own and his rival's payoffs and also that his rival has incomplete information; the other person knows only his own payoffs and does not know what information the first person has. These information-state distinctions continue to be of central importance in contemporary experimental studies.

A major result reported by SF is the strong tendency (given adequate motivation as noted above) for bargainers to negotiate Pareto-optimal

13. It is an empirical question whether complete information is equivalent behaviorally to common knowledge; in a cooperative bargaining context, Roth and Muringhan (1982) found no significant difference between complete information with, and without, common knowledge (also between incomplete information with, and without, common knowledge). Roth and Muringhan achieved common knowledge by instructing both subjects that they are both reading the same instructions. Somewhat more effective, presumably, is the common practice of achieving the latter by reading common instructions to all subjects simultaneously. It is an open empirical question whether these methods produce differences in behavior. Strictly interpreted, common knowledge means that you know it, I know it, we both know it, we both know that we both know it, and so on. If subjects are not explicitly told that they are or might be reading different instructions, it is perhaps likely that they assume that they are reading the same instructions.

contracts. This result continues to find strong support down to the present (see for example, Hoffman and Spitzer 1982 and 1985).

However, there has been a vast improvement over the years since SF in the sophistication applied to the study of cooperative bargaining behavior. For example, the work of SF did not attempt to test the Nash (1950) bargaining solution based on maximizing the product of the player's utility functions. This is because SF had no mechanism for controlling for the utility of bargaining outcomes. It was not until Roth (1979) that an ingenious solution to this problem was provided and then implemented experimentally by Roth and Malouf (1979). The procedure is simple: dyads bargain over the division of 100 identical lottery tickets each representing a chance to win a fixed large prize, or a fixed small prize, say M_i and m_i respectively for player i (the prizes are generally different for the two players). From von Neumann–Morgenstern utility theory, the utility of the bargain (lottery) is measured for each player by the percentage of the lottery tickets each receives in the final bargain.[14] (The small prize is also received if the bargainers fail to reach agreement.) Thus, if the bargain is 65 percent for one and 35 percent for the other, the players' utilities are respectively 0.65 and 0.35 for the bargaining outcome, with the utilities scaled so that $u_i\ (M_i) = 1$ and $u_i\ (m_i) = 0$. Roth and Malouf (1979) find strong support for the predictions of the Nash bargaining model in this context (a 50-50 split of the tickets; i.e., equal utilities) under incomplete information (whether or not this is common knowledge), and whether the prizes are equal or different for the two players. But under complete information, where both players know both *prizes* (whether or not this is common knowledge) the Nash prediction is not supported if the prizes differ between the two players. Thus the Nash solution performs best where payoff information is private. This is in precise accordance with Nash's theory: the axioms include the expected utility axioms, which by Roth's implementation means that the bargainers "know" each other's preferences; also Nash states explicitly (but not axiomatically) that "each (bargainer) has full knowledge of the tastes and preferences of the other" (1950, 155). Nash had no reason to suppose that the bargainers *knew the prizes* since the theory is stated

14. This lottery procedure had been proposed earlier by Cedric Smith (1961) as a means of controlling for risk-neutral behavior in games against nature. It was subsequently picked up and discussed by Savage (1971), which is the source from which economists were most likely to learn of it. I learned of it in correspondence with Savage when he sent me a prepublication draft (dated 25 May 1971) just before his death on 1 November 1971.

entirely in terms of outcomes in utility space, not commodity space. According to this interpretation, Nash makes no predictions in the case where the prizes are complete information. It is of course correct to say that the model predicts that the outcome depends only on the utilities, not the prizes, but only if those prizes are private information. In fact privacy is essential as a means of *controlling for interdependent utilities,* where the utility of either (or both) bargainer(s) depends upon the allocation of prize money to the other as well as to himself (Smith 1976a, 278; Smith 1982, 934–35).[15]

The SF experiments with complete-incomplete information constituted the first experimental examination of asymmetric information theory. Specifically they tested the Schelling (1957) hypothesis according to which the bargainer with less information may have an advantage over an adversary with complete information. The results support this hypothesis, but not significantly, probably because of the small samples (SF, 58). In recent research on asymmetric information, which reinforces this finding, the phenomenon has been called the "curse of knowledge" (see Camerer, Loewenstein, and Weber 1989, for new evidence in a market setting and for other references). Again, work by SF anticipated important subsequent research in experimental economics.

Noncooperative Bargaining and Oligopoly. In part 2 of their second book, FS again study bilateral bargaining: the case of unequal strength, by which they mean the price leadership model developed by Bowley (1928) who derived the noncooperative equilibrium for this case. Nash (1951; also see Luce and Raiffa 1957, 89) defines such games in terms of the absence of coalitions and of preplay communications. The theory assumes complete information and a single round of the game. The FS environment is like that in SF; that is, a buyer can redeem purchased units according to a specified (quadratic) revenue function, and the seller sells units off a (quadratic) cost function. The institution is posted pricing: the seller chooses and announces price to the buyer, and the buyer responds by choosing quantity. The Bowley-Nash equilibrium is at the price-quantity pair corresponding to a monopoly equilibrium. Five bargaining experiments are reported by FS: two use a single play with

15. The Nash axioms specifically rule out such interpersonal comparisons of utilities so that "when one side makes a concession to the other he will do so out of *self-interest,* that is because he thinks it is unlikely that he can reach an agreement . . . without making this concession" (Harsanyi 1987, 1:192). Under privacy it is possible for this Nash condition to be satisfied.

complete information; three are repeat play games for twenty periods whose length is unknown to the subjects.

Both of the single-play experiments provided strong support for the Nash equilibrium. In one experiment the equal split profit point corresponds to the Pareto-optimal allocation, and in the other this point corresponds to the Nash equilibrium. But the results were not significantly different from each other. Recently a class of bargaining games, called ultimatum games, have been studied experimentally. These games are identical in structure to the single-play posted price experiments of FS, if in the latter the buyer's choice of quantity is reduced to the $\{0, 1\}$ dichotomy—trade, or no trade at the posted price. Also, in an ultimatum game for the division, say, of $10, "the" Nash equilibrium is for the first mover to offer ϵ to his counterpart and retain $\$10 - \epsilon$ for himself. The problem here is that ϵ is not known a priori, since it is the smallest amount that the second mover will accept and allow the players to collect $\$10 - \epsilon$ and ϵ from the experimenter. (If the offer is rejected each gets zero.) Research has centered upon the question of whether the observed values of ϵ are "too large" to be consistent with the Nash equilibrium, and on the treatments that affect these observations (see Forsythe et al. 1988 for a summary and a report of new experiments). In the FS design (experiment 2), where the game is framed as a trade, if the seller offers the Nash price ($9) the buyer can respond with any nonnegative discrete quantity choice including the Nash quantity (10 units). At the Nash point profits are $6.44 for the seller, $2.44 for the buyer, which are the modal observed outcomes. It would appear that the problem with the ultimatum game, vis-à-vis FS, is that it has boundary problems (near-zero payoff for one player) [16] and/or it is presented as an abstract split-the-prize game rather than an exchange.[17]

16. See Mellers, Weiss, and Birnhaum (1990) for experimental studies identifying aberrant behavior at (or near) zero outcomes in evaluating gambles. In the ultimatum game, when the second mover rejects the offer as too small, the subject may simply be sending a message to the experimenter that he/she does not wish to be recruited for such unremunerative experiments in the future.

17. The interpretation is that if it is presented as a split-the-prize game, the first mover may feel that he/she has not really "earned" the property right to this asymmetric advantage and is not justified in exploiting it. But if it is described as an exchange, the person who is the seller may feel justified in pricing high. See Hoffman and Spitzer (1985) for a study of the effect of "earning" a property right as compared with having it assigned by an egalitarian coin flip. Further results strongly supporting this interpretation in ultimatum and dictator games are reported in Hoffman, McCabe, and Smith 1991.

In FS the three repeat-play games, two with complete information, one with incomplete information, support the following hypothesis: the Nash equilibrium receives its strongest support under incomplete information. In their Cournot quantity adjuster and Bertrand price adjuster oligopoly experiments Fouraker and Siegel again compare the incomplete and complete information treatments in the context of repeated trials. In oligopoly as in bargaining, the Nash noncooperative equilibrium receives its strongest support under incomplete information.[18] These results are contrary to conventional game-theoretic ideas in the following sense. Game theorists since Nash have supposed that complete information was necessary for a Nash equilibrium because its achievement was thought to be the result of conscious cognition and calculation, not the result of a trial-and-error groping process. (Players are supposed to arrive at a Nash equilibrium the way game theorists do.) Also, Nash defines an equilibrium for a single play, not repeated plays. In a repeated (super) game many equilibria are possible, all the way from the single-play Nash to the single-play monopoly, but traditionally this required complete information. However, Kreps, Milgrom, Roberts, and Wilson (1982) show that in the prisoner's dilemma game the cooperative outcome can be an equilibrium (until "near" the end) by introducing an ϵ-probability of incomplete information on the strategy of one's opponent. Their model, however, predicts that such players will come off the starting blocks at the cooperative outcome. As noted by Selten (1989, 22) this is not what experimental subjects do: if they achieve the cooperative outcome it is by a sequential learning process over time (for example, in Selten and Stocker 1986 and in FS). Across different institutions the FS results are mixed relative to the achievement of the cooperative outcome under complete information, but it is clear that private information is the condition most supportive of Nash. Furthermore, this finding has been supported in a great variety of experiments with institutions and environments different from those in FS (for a summary see McCabe, Rassenti, and Smith 1991). But new game theory research is addressing the problem of deducing theoretical results that are consistent with Nash

18. As noted by Shubik over three decades ago, "The *more information there is in a market, the more likely it is that combinations will result*" (1959, 172). The FS Cournot duopolies have been replicated recently by Moir and Mestelman (1991); their results confirm those of FS. Cournot behavior is more common under incomplete than complete information. Under complete information there are more responses that are collusive and more that are competitive than under incomplete information.

outcomes in repeated games of incomplete information (Canning 1990, Kalai and Lehrer 1990). This research is promising in providing better integration between game theory and experimentation.

The Middle Years, 1963–1975

In the spring semester 1963, experimental economics appeared to be developing sufficiently well that I ventured to offer my first workshop seminar in the subject at Purdue University (Economics 676). This seminar was repeated annually until 1967, when I went to Brown University. There was insufficient literature in experimental economics proper (the study of markets and of the behavior of individuals in the context of markets) for a full semester course. So in the beginning (1963–65), this course relied more heavily on the psychological and economic literature in individual decision behavior under uncertainty (for example, Siegel 1961; Edwards 1954, 1955, 1961; Ellsberg 1961; Dolbear 1963; Davidson, Suppes, and Siegel 1957) and some work in group decision (committee processes) (Wallach, Kogan, and Bem 1962). It also drew upon the considerable literature in matrix games, including prisoner's dilemma experiments, which existed in the early 1960s, much of it also due to psychologists (Scodel, Minas, Ratoosh, and Lipetz 1959; Minas, Scodel, Marlowe, and Rawson 1960; Lave 1965; Rapoport and Orwant 1962). The main economics studies were the ones already discussed above (Siegel 1961; SF 1960; FS 1963; Smith 1962 and 1964).

Although matrix games (using the classic work of Luce and Raiffa [1957]), and the study of choice behavior in such games, provided an elegant means of demonstrating equilibrium concepts and elements of conflict and cooperation in markets, I never viewed these austere environments as constituting the corpus of experimental economics. To reduce market competition, under the many institutional forms in which trade occurs, to a two-person (or larger) matrix of interdependent profit payoffs does much violence to the economics of exchange. It is a wise firm indeed that can identify its own cost, demand, and profit function, let alone reduce these, and the actions of a competitor to a bimatrix game of payoffs (see Shubik 1959, 17). Also the normal form game matrix trivializes the concept of an institution and the language of the market by reducing the distinctive effects of the environment and the institution to their net combined effect on payoffs.

It was in this period, 1963–67, that I worked out the theory of induced valuation (Smith 1976a), and, in conjunction with Donald Rice (Rice and Smith 1964), the beginning of an attempt to deal with questions of "parallelism" between laboratory and field (Smith 1980). By 1967 enough work had accumulated in experimental market economics that I had dispensed with most of the workshop's reading "filler" on matrix game and utility theory experiments. The new material included my lecture notes on induced valuation and parallelism, my new published or working paper studies (Smith 1965 and 1967, the latter a working paper from 1965–67); and several accumulated articles and working papers by past students in the seminar (Belovicz 1967; Johnson and Brennan 1963; Lipstadt 1966; Murphy 1966; and Williams 1973).

Beginning as early as 1957–58, and continuing through the 1960s, I gave occasional seminars on experimental economics, which was an avocation before 1973–74 and thereafter—my vocation being investment and capital theory, later the economics of uncertainty, finance, and natural resource economics. But audience reactions to presentations in experimental economics differed dramatically from the reactions to papers using more familiar economic techniques and tools. The experimental papers tended to elicit methodological questions. There must be something wrong with these experiments? How can you get competitive outcomes in the absence of complete information? Experiments should only be used to test previously well-formulated theory. What do you expect to learn from students playing market games for low-stakes payoffs? And so on. I was constantly pushed to justify why I did what I did and how I did it. Seminars on any other research topic that I found attractive never elicited questions that in effect challenged me to explain why I was doing that research. I puzzled about this, and this contemplation drew me more and more into methodology, but it tended to be methodology done in the context of particular examples and research programs.

In late 1963, thanks to the initiative of Dick Cyert at what was then Carnegie Tech, Lester Lave and I organized a Faculty Research Workshop in Experimental Economics sponsored by the Ford Foundation, which was given in the summer of 1964 at Carnegie. Response was positive, and we decided to repeat the workshop with a different group in the summer of 1965. The participants included John Carlson, Trenery Dolbear, Bartel Jensen, Austin Hoggatt, Gordon Sherman, William Starbuck, Richard Swensson, and many others. Lester Lave and I attempted to find a publisher for a selection of refereed papers that came out of these two summer workshops, but we were unsuccessful. We gave up in early 1966 and released all of the authors to submit their papers to journals. There was simply no market for such a volume at that time. (The McGraw-Hill editor confessed that FS 1963 had sold only 2100 copies by that time, and they could not take on any more of such manuscripts.)

In 1969 Michael J. Farrell, editor of the *Review of Economic Studies*, sponsored the publication of a Symposium on Experimental Economics. Also in 1969, Jim Friedman and I attempted to find a publisher for a collection of reprints in experimental economics, since there existed by then many published papers from which one could select for such a volume. Again, there was simply not an adequate market to justify such an effort.

From 1968 to 1973, I was not active in experimental research, although I continued to maintain an interest in it. (My primary research output was in the theoretical areas of natural resource economics and the economics of uncertainty.) But of great significance to me during this period was that Charles Plott and I spent countless hours discussing experimental (and other) economics while bass fishing in the canyons of Utah's Lake Powell. Charlie became fascinated with experimental economics, and these discussions were easily the most important influence on me since my encounter with Sid Siegel in 1961. Then in 1972–73 I became a visiting Fellow at the Center for Advanced Study in the Behavioral Sciences. There I completed various projects in the economics of uncertainty and natural resources, and I was ready to return to experimental economics. Charlie was instrumental in getting me to come to Cal Tech as a Fairchild Scholar, 1973–74 (along with Bill Riker, also a Fairchild Scholar that year). In some ways the most important aspect of that year is that my collaboration with Charles Plott induced me to get into collaborative research as a way of experimental life. Experimental research, far more than any other type of economics research, calls for team research. With Charlie I learned the enormous synergy that could come out of team effort. We discovered together a new significance for institutions, which I had never fully appreciated while working alone; yet that was what I had been intimately involved with in Smith 1962, 1964, and 1967. This led to our first paper (Plott and Smith 1978), but the result of that joint learning is evident in many later works (for example, Smith 1982 and Plott 1986).

At Cal Tech, in the spring quarter of 1974, Charlie and I team-taught a seminar in experimental economics. I dusted off my old reading list and lecture notes from Purdue, circa 1967; we updated it; and we had a delightful experience with perhaps three paying student customers and such active faculty participants as John Ferejohn, Mo Fiorina, Roger Noll, Bill Riker, and Jim Quirk. This led to the Cal Tech program in social science (both undergraduate and graduate) which made extensive use of classroom experiments. I moved to the University of Arizona in 1975 and helped plant similar traditions there.

After 1975 experimental economics experienced such an explosion of growth that a separate and larger-scale treatment is essential and well beyond my capacity to accomplish at this time.

References

Aumann, R. J. Game Theory. 1987. In *The New Palgrave*, vol. 2, edited by J. Eatwell, M. Milgate, and P. Newman. London: Macmillan.

Bellman, R., et al. 1957. On the Construction of a Multistage, Multiperson Business Game. *Journal of the Operations Research Society of America* 5 (August): 469–503.

Belovicz, Meyer W. 1967. The Sealed Bid Auction: Experimental Studies. Ph.D. thesis, Purdue University.

Bowley, A. L. 1928. On Bilateral Monopoly. *The Economic Journal* 38 (December): 651–59.

Camerer, Colin, George Loewenstein, and Martin Weber. 1989. The Curse of Knowledge in Economic Settings: Experimental Analysis. *Journal of Political Economy* 97.5 (October): 1232–54.

Canning, David. 1990. Convergence to Equilibrium in a Sequence of Games with Learning. Discussion Paper. London School of Economics.

Chamberlin, Edward. 1948. An Experimental Imperfect Market. *Journal of Political Economy* 56 (April): 95–108.

Coppinger, Vicky, Vernon L. Smith, and Jon A. Titus. 1980. Incentives and Behavior in English, Dutch, and Sealed Bid Auctions. *Economic Inquiry* 18 (January): 1–22.

Davidson, D., P. Suppes, and S. Siegel. 1957. *Decision Making: An Experimental Approach*. Stanford: Stanford University Press.

Dolbear, F. T. 1963. Individual Choice under Certainty: An Experimental Study. *Yale Economic Essays* 3.2:419–70.

Edwards, Ward. 1954. Theory of Decision Making. *Psychological Bulletin* 51.4 (July): 380–417.

———. 1955. The Prediction of Decision among Bets. *Journal of Experimental Psychology* 50.3 (September): 204–14.

———. 1961. Behavioral Decision Theory. In *Annual Review of Psychology*, vol. 12, edited by P. R. Farnsworth.

Ellsberg, D. 1961. Risk, Ambiguity, and the Savage Axioms. *Quarterly Journal of Economics* 75.4 (November): 643–69.

Forsythe, Robert, J. L. Horowitz, N. E. Savin, and M. Sefton. 1988. Replicability, Fairness, and Pay in Experiments with Simple Bargaining Games. University of Iowa, Working Paper 88-30, December.

Fouraker, L. E., and Sidney Siegel. 1961. *Bargaining Behavior II*. Department of Psychology, Pennsylvania State University.

Fouraker, Lawrence, and Sidney Siegel. 1963. *Bargaining Behavior*. New York: McGraw-Hill. (Referred to in text as FS.)

Fouraker, Lawrence, M. Shubik, and S. Siegel. 1961. Oligopoly Bargaining: The Quantity Adjuster Models. *Research Bulletin* 20. Department of Psychology, Pennsylvania State University.

Fouraker, L. E., Sidney Siegel, and Donald Harnett. 1961. *Bargaining Behavior I*. Department of Psychology, Pennsylvania State University.

Friedman, James. 1963. Individual Behavior in Oligopolistic Markets. *Yale Economic Essays* 3.2:359–417.

———. 1967. An Experimental Study of Cooperative Duopoly. *Econometrica* 35 (October): 379–97.

Friedman, James, and Austin Hoggatt. 1980. An Experiment in Noncooperative Oligopoly. Supplement 1. V. L. Smith, ed.

Harrison, Glenn, 1989. Theory and Misbehavior in First Price Auctions. *American Economic Review* 79 (September): 749–62.

Harsanyi, John. 1986. Bargaining. In *The New Palgrave*, vol. 1, edited by J. Eatwell, M. Milgate and P. Newman. London: Macmillan.

Hoffman, Elizabeth, and Matthew L. Spitzer. 1982. The Coase Theorem: Some Experimental Tests. *Journal of Law and Economics* 25.1 (April): 73–98.

———. 1985. Entitlements, Rights, and Fairness: An Experimental Examination of Subjects' Concepts of Distributive Justice. *Journal of Legal Studies* 14 (June): 259–97.

Hoffman, Elizabeth, Kevin A. McCabe, and Vernon L. Smith. 1991. Fairness, Property Rights, and Bargaining. Photocopy. University of Arizona.

Hoggatt, Austin C. 1959. An Experimental Business Game. *Behavioral Science* 4 (July): 192–203.

Johnson, Clarke C., and R. D. Brennen. 1963. The Prisoner's Dilemma Game: Some Empirical and Theoretical Proposals. Paper No. 42. Institute for Quantitative Research in Economics and Management, Purdue University.

Kalai, Ehud, and E. Lehrer. 1990. Rational Learning Leads to Nash Equilibrium. Discussion Paper No. 895. Kellogg Graduate School of Management, Northwestern University.

Kreps, D., et al. 1982. Rational Cooperation in the Finitely Repeated Prisoner's Dilemma. *Journal of Economic Theory* 27.2:245–53.

Lave, Lester. 1965. Factors Affecting Co-operation in the Prisoner's Dilemma. *Behavioral Science* 10.1 (January): 26–38.

Leffler, G., and C. L. Farwell. 1949. *The Stock Market*. New York: Macmillan.

Lewis, D. K. 1969. *Convention*. Cambridge, Mass.: Harvard University Press.

Lipstadt, Stanley. 1966. Wherefor Art Thou, Punishing Bid? Face-to-Face Bargaining in Bilateral Monopoly. Mimeo. Purdue University.

Luce, Duncan, and Howard Raiffa. 1957. *Games and Decisions*. New York: John Wiley.

McCabe, Kevin A., Stephen J. Rassenti, and Vernon L. Smith. 1991. Lakatos and Experimental Economics. In *Appraising Economic Theories*, edited by N. de Marchi and M. Blaug. Aldershot: Edward Elgar.

Mellers, Barbara, Robin Weiss, and Michael Birnhaum. 1990. Violations of Dominance in Pricing Judgements. Photocopy. Department of Psychology, University of California.

Messick, Samuel, and A. H. Brayfield. 1964. *Decision and Choice*. New York: McGraw-Hill.

Minas, J. A., D. Marlowe, A. Scodel, and H. Rawson. 1960. Some Descriptive Aspects of Two-Person Non-Zero-Sum Games I. *The Journal of Conflict Resolution* 4.2 (June): 193–97.

Mirowski, Philip. 1991. When Games Grow Deadly Serious: The Military Influence on the Evolution of Game Theory. In *Economics and National Security*, edited by Craufurd Goodwin. Durham, N.C.: Duke University Press.

Moir, Robert, and Stuart Mestelman. 1991. Does Cournot Live in the Lab? Photocopy. Department of Economics, McMasters University.

Murphy, James. 1966. Effects of the Threat of Losses on Duopoly Bargaining. *Quarterly Journal of Economics* 80 (May): 296–313.

Nash, John. 1950. The Bargaining Problem. *Econometrica* 18 (April): 155–62.

———. 1951. Noncooperative Games. *Annals of Mathematics* 54 (September): 286–95.

Newell, Allen, and Herbert Simon. 1972. *Human Problem Solving*. New York: Prentice Hall.

Plott, Charles R., and Vernon L. Smith. 1978. An Experimental Examination of Two Exchange Institutions. *Review of Economic Studies* 45 (February): 133–53.

———. 1986. Laboratory Experiments in Economics. *Science* 232:732.

Rapoport, Anatol, and Carol Orwant. 1962. Experimental Games: A Review. *Behavioral Science* 7 (January): 1–37.

Rice, Donald B., and Vernon L. Smith. 1964. Nature, the Experimental Laboratory, and the Credibility of Hypotheses. *Behavioral Science* 9 (July): 239–46.

Roth, Alvin. 1979. Axiomatic Models of Bargaining. Lecture Notes in Economics and Mathematical Systems, No. 170. Berlin: Springer-Verlag.

Roth, Alvin, and M. K. Malouf. 1979. Game-Theoretic Models and the Role of Information in Bargaining. *Psychological Review* 86:1123–42.

Roth, Alvin, and J. K. Murnighan. 1982. The Role of Information in Bargaining: An Experimental Study. *Econometrica* 50.5 (September): 1123–42.

Sauermann, H., ed. 1967–72. *Contributions to Experimental Economics*, vols. 1–3. Tübingen: Mohr.

Sauermann, H., and Reinhard Selten. 1960. An Experiment in Oligopoly. In *General Systems Yearbook of the Society for General Systems Research*, vol. 5. Edited by L. Bertalarffy and A. Rapoport. Ann Arbor, Mich.: Society for General Systems Research.

Savage, Leonard J. 1971. Elicitation of Personal Probabilities and Expectations. *Journal of the American Statistical Association* 66 (December): 783–801.

Schelling, T. C. 1957. Bargaining, Communication, and Limited War. *Journal of Conflict Resolution* 1.1 (March): 19–36.

Scodel, A., V. Minas, P. Ratoosh, and M. Lipetz. 1959. Some Descriptive Aspects of Two-Person Non-Zero-Sum Games, I. *The Journal of Conflict Resolution* 3.2 (June): 114–19.

Selten, Reinhard. [1989] 1990. Evolution, Learning, and Economic Behavior. Nancy Schwartz Lecture. Kellogg Graduate School of Management, Northwestern University.

Selten, Reinhard, and Rolf Stoecker. 1986. End Behavior in Sequences of Finite Prisoner's Dilemma Supergames. *Journal of Economic Behavior and Organization* 7 (March): 47–70.

Shubik, Martin. 1959. *Strategy and Market Structure*. New York: John Wiley.

———. 1962. Some Experimental Non-Zero-Sum Games with Lack of Information about the Rules. *Management Science* 8.2 (January): 215–34.

———. 1970. A Note on a Simulated Stock Market. *Decision Sciences* 1.1 (January): 129–41.

———. 1975. *The Uses and Methods of Gaming*. New York: Elsevier Scientific.

Shubik, M., G. Wolf, and H. Eisenberg. 1972. Some Experiences with an Experimental Oligopoly Business Game. *General Systems* 13:61–75.

Shubik, M., and Richard Levitan. 1980. *Market Structure and Behavior*. Cambridge, Mass.: Harvard University Press.

Siegel, Sidney. 1959. Theoretical Models of Choice and Strategy Behavior: Stable-State Behavior in the Two-Choice Uncertain Outcomes Situation. *Psychometrika* 24.4 (December): 303–16.

———. 1961. Decision Making and Learning under Varying Conditions of Reinforcement. *Annals of the New York Academy of Science* 89:766–83.

Siegel, Sidney, and Donald Harnett. 1961. Bargaining, Information, and the Use of Threat. Research Bulletin 21. Pennsylvania State University, Department of Psychology.

Siegel, Sidney, and L. E. Fouraker. 1960. *Bargaining and Group Decision Making*. New York: McGraw-Hill. (Referred to in text as SF.)

Simon, Herbert A. 1955. A Behavioral Model of Rational Choice. *Quarterly Journal of Economics* 69 (February): 99–118.

———. 1956a. A Comparison of Game Theory and Learning Theory. *Psychometrika* 21 (September): 267–72.

———. 1956b. Rational Choice and the Structure of the Environment. *Psychological Review* 63 (March): 129–38.

Smith, Cedric A. B. 1961. Consistency in Statistical Inference and Decision. *Journal of the Royal Statistical Society*, series B. 23.1:1–25.

Smith, Vernon L. 1962. An Experimental Study of Competitive Market Behavior. *Journal of Political Economy* 70 (April): 111–37.

———. 1964. Effect of Market Organization on Competitive Equilibrium. *Quarterly Journal of Economics* 78 (May): 181–201.

———. 1965. Experimental Auction Markets and the Walrasian Hypothesis. *Journal of Political Economy* 73 (August): 387–93.

———. 1967. Experimental Studies of Discrimination versus Competition in Sealed-Bid Auction Markets. *Journal of Business* 40 (January): 56–84.

———. 1976a. Experimental Economics: Induced Value Theory. *American Economic Review* 66 (May): 274–79.

———. 1976b. Bidding and Auctioning Institutions: Experimental Results. In *Bidding and Auctioning for Procurement and Allocation*, edited by Y. Amihud. New York: New York University Press.

———. 1980. Relevance of Laboratory Experiments to Testing Resource Allocation Theory. In *Evaluation of Econometric Models*, edited by J. Kmenta and J. R. Ramsey. New York: Academic.

———. 1981. Experimental Economics at Purdue. In *Essays in Contemporary Fields of Economics*, edited by G. Horwich and J. P. Quirk. West Lafayette, Ind.: Purdue University Press.

———. 1982. Microeconomics Systems as an Experimental Science. *American Economic Review* 72 (December): 923–55.

———. 1991. Rational Choice: The Contrast Between Economics and Psychology. *Journal of Political Economy* 99 (August): 877–97.

Smith, Vernon L., and James M. Walker. 1990. Monetary Rewards and Decision Costs in Experimental Economics. Economic Science Laboratory, University of Arizona (to appear in *Economic Inquiry*).

Stigler, George J. 1968. Competition. In *International Encyclopedia of the Social Sciences*, vol. 3. New York: Macmillan.

Thaler, Richard. 1986. The Psychology and Economics Handbook: Comments on Simon, on Einhorn and Hogarth, and on Tversky and Kahneman. In *Rational Choice*, edited by R. Hogarth and M. Reder. Chicago: University of Chicago Press.

Thrall, R. M., C. H. Coombs, and R. L. Davis. 1954. *Decision Processes*. New York: John Wiley. (Referred to in text as TCD.)

Thurston, L. L. 1931. The Indifference Function. *Journal of Social Psychology* 2 (May): 139–67.

Tversky, Amos, and Ward Edwards. 1966. Information versus Reward in Binary Choice. *Journal of Experimental Psychology* 71.5 (May): 680–83.

Tversky, Amos, and Daniel Kahneman. 1986. Rational Choice and the Framing of Decision. In Hogarth and Reder 1986.

Wallach, M. A., N. Kogan, and D. V. Bem. 1962. Group Influence on Individual Risk Taking. *Journal of Abnormal Social Psychology* 65.2 (August): 75–86.

Wallis, W. A., and Milton Friedman. 1942. The Empirical Derivation of Indifference Functions. In *Studies in Mathematical Economics and Econometrics in Memory of Henry Schultz*, edited by O. Langs, F. McIntyre, and T. O. Yntema. Chicago: University of Chicago Press.

Williams, Fred. 1973. Effect of Market Organization on Competitive Equilibrium: The Multiunit Case. *Review of Economic Studies* 40 (January): 97–113.

Index

Contributors

Mary Ann Dimand lectures on economics at Brock University, Canada, and is the author (with Robert Dimand) of *Trade, Uncertainty, and Commodity Storage* (1992).

Robert W. Dimand is professor of economics at Brock University, Canada, and is the author of *The Origins of the Keynesian Revolution* (1988).

Robert J. Leonard is assistant professor of economics at University of Québec at Montréal. He is interested in the history of contemporary economic theory.

Philip Mirowski is Carl Koch Professor of Economics and the History and Philosophy of Science at the University of Notre Dame. His books include *More Heat than Light* (1989), *Against Mechanism* (1988), and a forthcoming edited volume on natural images in economics, *Markets Read in Tooth and Claw*.

Angela M. O'Rand is associate professor of sociology at Duke University. She is author of *Disciples of the Cell* (forthcoming), a study of scientific development in a specialty of modern biology.

Howard Raiffa is the Frank Plumpton Ramsey Professor of Managerial Economics at the Graduate School of Business Administration, Harvard University.

Urs Rellstab is a Ph.D. student at the University of St. Gallen, Switzerland. With support of the Swiss National Science Foundation he spent a year (1990–91) at Duke University where he mainly worked with the Oskar Morgenstern Papers.

Robin E. Rider is head of the History of Science and Technology Program at the Bancroft Library, University of California at Berkeley. Her recent research projects include the history of operations research and the history of the printing of science.

William H. Riker is Wilson Professor of Political Science, Emeritus, at the University of Rochester. He is the author of *The Theory of Political Coalitions*, *Liberalism against Populism*, *The Art of Political Manipulation*, and many papers on formal political theory.

Andrew Schotter is professor and chair of the department of economics at New York University. He is the author of *The Economic Theory of Social Institutions* (1981) and *Free Market Economics: A Critical Appraisal* (1987).

Martin Shubik is the Seymour Knox Professor of Mathematical Institutional Economics at Yale University. He is the author of several books including *Strategy and Market Structure* (1959) and *Game Theory in the Social Sciences* (1982). He is currently working on a book on the theory of money and financial institutions.

Vernon L. Smith is Regents' Professor of Economics and Director of Research, Economic Science Laboratory, University of Arizona. He is a Fellow of the Econometric Society, the American Association for the Advancement of Science, and the American Academy of Arts and Sciences.

E. Roy Weintraub is professor of economics at Duke University. Trained as a mathematician, he currently writes on the history of the interrelationship between economics and mathematics. Author of a half-dozen books and numerous articles, he offers one of the first constructivist accounts of economic analysis in his most recent book, *Stabilizing Dynamics* (1991).